H. M. Dietz V. Nollau

Markov Decision Problems with Countable State Spaces

Mathematical Research · Mathematische Forschung

Wissenschaftliche Beiträge
herausgegeben von der
Akademie der Wissenschaften der DDR
Institut für Mathematik

Band 15

Markov Decision Problems with Countable State Spaces

by

H.M. Dietz
V. Nollau

Markov Decision Problems with Countable State Spaces

Optimality Criteria - Algorithms - Clustering

by

Hans Michael Dietz

and

Volker Nollau

Akademie-Verlag · Berlin 1983

Verfasser:
Dr. rer.nat. Hans Michael Dietz
Doz. Dr. sc.nat. Volker Nollau
Technische Universität Dresden, Sektion Mathematik
8027 Dresden, Mommsenstr. 13

Die Titel dieser Schriftenreihe werden vom Originalmanuskript der Autoren reproduziert.

Erschienen im Akademie-Verlag, DDR-1086 Berlin, Leipziger Str. 3-4
Lektor: Dr. Reinhard Höppner

Lizenznummer: 202 • 100/411/83
Gesamtherstellung: VEB Kongreß- und Werbedruck, 9273 Oberlungwitz
Umschlag: Rolf Kunze
Bestellnummer: 763 224 0 (2182/15) • LSV 1075
Printed in GDR
DDR 22, – M

INTRODUCTION

When Richard BELLMAN published his famous book "Dynamic Programming" in the 1950's, it was the beginning of a very new mathematical branch of knowledge. A main part of this rapidly developing theory is "Stochastic Dynamic Programming" which deals with the mathematical research of multistage stochastic decision processes. In this theory one takes account of the double role of the decisions at each step: the immediate pay-off and the stochastic influence on the future evolution of the considered system. Already in BELLMAN's first monograph (1957) the stochastic case was considered along with the deterministic case. Further essential contributions in this field were made by HOWARD [60], BLACKWELL [9], STRAUCH [108] and HINDERER [49].

The present study based on the thesis of H.-M. DIETZ [24] is devoted to some topics of contemporary theory of special multistage stochastic Decision Processes known as Markov Decision Processes.

In the first chapter of this book we introduce some basic definitions and results concerning Markov chains with countable state space. We consider the DOEBLIN condition with respect to a transition matrix and the characterization of this condition by the quasi-compactness of the related transition operator. Some essential conclusions from this characterization were obtained by methods of spectral theory of linear operators. The section 1.3 contains some considerations concerning the clustering of the state space of Markov chains introduced by ROSENBLATT [93] and KEMENY and SNELL [71].

The second chapter deals with the theory of Markov Decision Problems (MDP's) with countable state space. Firstly, we consider MDP's with finite horizon and discounted MDP's, which are based on general models investigated by HINDERER [49], SCHÄL [101], BLACKWELL [8], DERMAN [19], MAITRA [80], [81], HARRISON [43], FURUKAWA [37], and WESSELS [119]. The following sections deal with the average reward optimality and generalizations of the HOWARD algorithm for countable state space also considered by DERMAN [18], KUSHNER [75], IWAMOTO [63], KÜENLE [74], and ASO and KIMURA [1] in the single-chain case. In section 2.5 and 2.6 we generalize some results of MILLER and VEINOTT [84], HORDIJK and SLADKY [58], and KADOTA [66] concerning sensitive optimization.

In recent years some algorithms for state space clustering in MDP's have been considered by several authors (BERTSEKAS [7], FOX [35], [36], HAHNEWALD-BUSCH and NOLLAU [41], [42], [86] (cf. [40])).

In the third chapter we investigate a new clustering procedure for all models considered in the preceding chapter. Thereby, a main result concerns the parallelity of the HOWARD-algorithm with respect to the given model and the clustered one. Moreover, we discuss possibilities and bounds of such a cluster procedure.

The appendix is devoted to the theory of infinite matrices and their related operators. Finally, the reader will find a summary of the symbols used in this study at the end of the book.

We thank Mrs. G. Böhm for her really excellent typing. And finally we thank the Akademie-Verlag for letting us publish this monograph in its series "Mathematical Research".

FSC
www.fsc.org
MIX
Papier | Fördert gute Waldnutzung
FSC® C083411

Kolibri 360 GmbH
Ferdinand-Jühlke-Straße 7
99095 Erfurt, Deutschland
produktsicherheit@kolibri360.de

CONTENTS

1. MARKOV CHAINS WITH COUNTABLE STATE SPACE

1.1. DEFINITIONS AND BASIC PROPERTIES

Let $(\Omega, \mathfrak{A})$ and $(E, \mathfrak{E})$ be measurable spaces.

Definition 1.1. A family $(X_n)_{n\in\mathbb{N}_0}$ of $(\mathfrak{A}, \mathfrak{E})$-measurable mappings $X_n: \Omega \to E$ is called a process with values in E. If $(\Omega, \mathfrak{A})$ is supplied with a probability measure, then $(X_n)_{n\in\mathbb{N}_0}$ is called a stochastic process on the probability space $(\Omega, \mathfrak{A}, P)$. Then the set E is called the state space of $(X_n)_{n\in\mathbb{N}_0}$, which we shall denote by (X_n) in the following.

Definition 1.2. A stochastic process (X_n) on $(\Omega, \mathfrak{A}, P)$ with values in E is called Markovian if for all $n \in \mathbb{N}_0$, $m \in \mathbb{N}$ and $\Gamma \in \mathfrak{E}$ $P(X_{n+m} \in \Gamma | X_n, X_{n-1}, \ldots, X_0) = P(X_{n+m} \in \Gamma | X_n)$ P-a.s. holds. (X_n) is called a Markov chain if E is countable.

Hereafter, we consider Markov chains, i.e., the state space is generally supposed to be countable. Unless otherwise stated, we assume that $\mathfrak{E} = \mathfrak{P}(E)$.

Definition 1.3. A real E-row vector $q \geq 0$ is called substochastic if it satisfies:

$$\sum_{x\in E} q(x) \leq 1.$$

For $\sum_{x\in E} q(x) = 1$ it is called stochastic. The set of all such vectors is denoted by $\mathfrak{M}^1(E)$. An $E \times E$-matrix is named substochastic or stochastic if its rows are substochastic and stochastic, respectively. A stochastic matrix is also called a transition matrix (on E).

It is obvious that the collection of substochastic matrices represents a subset of $\mathfrak{R}(E)$ (set of all $E \times E$-matrices M with $\sup_{x\in E} \sum_{y\in E} |M(x,y)| < \infty$). Therefore, the product PQ of two substochastic matrices P and Q is well-defined and, of course, a substochastic matrix, too. If both P and Q are stochastic, then so is the product matrix PQ.

Definition 1.4. A family $(Q_{m,n})_{m\in\mathbb{N}_0, m \leq n}$ of transition matrices on E is called a transition function if, for all $l \leq m \leq n$, $l \in \mathbb{N}_0$, the equation

$$Q_{l,n} = Q_{l,m}\, Q_{m,n}$$

is valid.

This definition differs slightly from that often used in literature. However, it seems to be justified to the authors because in the theory of Markov chains measure-theoretical considerations have no significance.

The connection between transition functions and Markov chains is based on the following

Definition 1.5. Let (X_n) be a Markov chain on the probability space $(\Omega, \mathfrak{A}, P)$ with the state space E and $(Q_{m,n})$ a transition function on E. $(Q_{m,n})$ is said to be the transition function of (X_n) if, for all $m,n \in \mathbb{N}_0$ with $m \leq n$ and all $y \in E$, the relation

$$P(X_n = y | X_m) = Q_{m,n}(X_m, y)$$

is fulfilled P-a.s.

If $(Q_{m,n})$ is a transition function on E and $q \in \mathfrak{M}^1(E)$, then there exists a Markov chain (X_n) with the transition function $(Q_{m,n})$ and the initial distribution given by q. Such a Markov chain can be constructed in the following way: One defines $A := E^{\mathbb{N}_0}$, $\mathfrak{A} = \bigotimes_{\mathbb{N}_0} \mathfrak{P}(E)$ and considers the process $X_n: A \to E$ given by the canonical projections

$$\omega = (x_0, x_1, \ldots, x_n, \ldots) \to X_n(\omega) := x_n \quad (n \in \mathbb{N}_0).$$

In the following we call $(A, \mathfrak{A})$ the canonical basic space and (X_n) the canonical process. For every cylinder set

$$[X_0 = x_0, \ldots, X_n = x_n] := \{\omega \in A: X_0(\omega) = x_0, \ldots, X_n(\omega) = x_n\}$$

we define

$$\mathbb{P}_q(X_0 = x_0, X_1 = x_1, \ldots, X_n = x_n) :=$$
$$q(x_0)Q_{0,1}(x_0, x_1) \cdot \ldots \cdot Q_{n-1,n}(x_{n-1}, x_n).$$

As is well-known, the σ-algebra $\mathfrak{A}$ is generated by the system of all such cylinder sets. Following a theorem of IONESCU-TULCEA (cf. [85]) the measure $\mathbb{P}_q$ has a unique extension on $\mathfrak{A}$. This probability measure will also be denoted by $\mathbb{P}_q$, where the index q marks the dependence on the chosen initial distribution. Due to this construction (X_n) becomes a Markov chain on $(A, \mathfrak{A}, \mathbb{P}_q)$ with $(Q_{m,n})$ as the transition function and the initial distribution q:

$$\mathbb{P}_q(X_0 = x_0) = q(x_0), \; x_0 \in E.$$

In particular, we are interested in Markov chains whose transition functions have a simple structure.

Definition 1.6. A transition function $(Q_{m,n})$ on E is called homogeneous if there exists a family $(P_n)_{n \in \mathbb{N}_0}$ of transition functions with

$$Q_{m,n} = P_{n-m}$$

for all $m,n \in \mathbb{N}_0$ and $m < n$. A Markov chain is called homogeneous if it possesses a homogeneous transition function.

For the family $(P_n) = (P_n)_{n \in \mathbb{N}_0}$ just introduced the relation

$$P_{m+n} = P_m P_n \quad (m,n \in \mathbb{N}_0) \qquad (1.1)$$

is evident. Beyond it we shall assume

$$P_0 = I.$$

Hence (P_n) forms a semi-group of stochastic matrices, and it holds that

$$P_n = P_1^n,$$

i.e., the semi-group (P_n) is generated by the (one-step) transition matrix $P := P_1$. Thus, in the following, we identify every homogeneous transition function with its transition matrix.

Obviously, the behaviourof a Markov chain is essentially determined by its transition matrix. In this section we give a survey on the main properties of a Markov chain which result from the structure of the transition matrix and which can be expressed by the matrix calculus.

Since an arbitrary transition matrix can be identified with the corresponding bounded linear operator on B(E) [1]), in general, it is not necessary to distinguish between the matrix P and the linear operator P. If such a dinstinction is necessary, we avoid confusion by a suitable formulation.

In the following, let P be a transition matrix on a countable set. Firstly we continue with some definitions. Let x and y be two states. Then y is called a successor of x if there exists an $n \geq 0$ for which $(P^n)(x,y) > 0$. In this case we write $x \rightsquigarrow y$. If $x \rightsquigarrow y$ and $y \rightsquigarrow x$ we say that x and y communicate. This relation decomposes the states of E into equivalence classes. A transition matrix is called irreducible if all states communicate. A subset $A \subset E$ is said to be closed if, for all $x \in A$,

$$\sum_{y \in A} P(x,y) = 1$$

holds. A closed subset is minimal if it does not contain any smaller closed subset.

For a state x the greatest common divisor of all integers $n \in \mathbb{N}$ with $(P^n)(x,x) > 0$ is called the period of x [2]). All states of an equivalence class have the same period (cf. [11]). The transition matrix P is said to be non-cyclic if the period of its states is one or infinite.

For every state $x \in E$ the symbol ε_x means the probability measure on $(E, \mathfrak{P}(E))$ concentrated in the point x, and we denote $\mathbb{P}_x := \mathbb{P}_{\varepsilon_x}$. A state $x \in E$ is said to be recurrent if

---

[1]) B(E) - set of all bounded, complex - valued functions on E (cf. appendix).

[2]) If there is no integer $n \in \mathbb{N}$ satisfying $(P^n)(x,x) > 0$, the period is set ∞.

$$\mathbb{P}_x(X_n = x_n \text{ for some } n \in \mathbb{N}) = 1,$$

otherwise x is said to be transient. Roughly speaking a recurrent state is a state to which the chain will almost surely return if it started there.

Lemma 1.1. [11]

(i) The states of an equivalence class are either all recurrent or all transient. A class of recurrent states forms a minimal closed subset.

(ii) A state $x \in E$ is transient, if and only if

$$\sum_{n=0}^{\infty} P^n(x,y) < \infty.$$

Based on this proposition we denote a class of recurrent states, in short, a recurrence class.

For every state $y \in E$ let τ_y be the function defined by

$$\tau_y(\omega) := \inf\{n \in \mathbb{N}: X_n = y\}$$

if $\inf\{n \in \mathbb{N}: X_n(\omega) = y\}$ is finite, otherwise

$$\tau_y(\omega) := \infty.$$

This function $\tau_y: A \to \mathbb{N} \cup \{\infty\}$ is evidently $(\mathcal{A}, \mathfrak{P}(N \cup \{\infty\}))$-measurable. It is easy to see that a state x is recurrent, if and only if $\mathbb{P}_x(\tau_x < \infty) = 1$. If $\mathbb{E}_x$ denotes the operator corresponding to $\mathbb{P}_x$, then a recurrent state x is said to have ergodic degree n ($n \in \mathbb{N}_0$) if $\mathbb{E}_x \tau_x^n < \infty$ and $\mathbb{E}_x \tau_x^{n+1} = \infty$. If such an integer n does not exist we define the ergodic degree to be equal to ∞.

Lemma 1.2. [72] All states of a recurrence class have the same ergodic degree.

A recurrent state is said to be null-recurrent if the ergodic degree is zero, and positive-recurrent if the ergodic degree is at least one. A recurrent class whose states have, at least, the ergodic degree two is said to be strongly ergodic.

Following the preceding considerations the state space E of a homogeneous Markov chain with the transition matrix P can be decomposed in the following way:

$$E = R \cup T,$$

where R and T are the sets of recurrent and transient states, respectively. Both sets can be empty. If there are some recurrent states, then they are either communicated or the set R consists of a countable number of recurrence classes $R_1, R_2, \ldots$

With respect to the properties of P^n for n tending to infinite one obtains the

Theorem 1.3.

(i) For all x, y $\in$ E there is the limit

$$P^*(x,y) := \lim_{N\to\infty} \frac{1}{N} \sum_{n=0}^{N-1} P^n(x,y).$$

(ii) The state y is positive recurrent and $x \rightsquigarrow y$, if and only if $P^*(x,y) > 0$.

(iii) Let $y \in E$ be a positive-recurrent state and R_1 the recurrent class containing y. Then, for any $x, x' \in R_1$,

$$P^*(x,y) = P^*(x',y) =: \pi(y),$$

and for any $x \in T$

$$P^*(x,y) = f^*(x, R_1)\pi(y)$$

with

$$f^*(x, R_1) = \mathbb{P}_x(X_n \in R_1 \text{ for some } n \in \mathbb{N}).$$

(iv) The series $\sum_{y\in E} (\frac{1}{N} \sum_{n=0}^{N-1} P^n(x, y))$ converges uniformly with respect to N for any $x \in E$, if and only if $\sum_{y\in E} P^*(x, y) = 1$.

(v) The matrix P^* with the elements $P^*(x, y)$ defined above satisfies the equations

$$P^*P = PP^* = (P^*)^2 = P^*.$$

Proof:

(i) - (iv): [11], I. 6.4. and I. 7.3.

(v): $P^*P = P^*$: Using FATOU's lemma one obtains

$$\begin{aligned}(P^*P)(x, y) &= \sum_{z\in E} P^*(x, z)P(x, y)\\ &= \sum_{z\in E} \lim_{N\to\infty} \frac{1}{N} \sum_{n=0}^{N-1} P^n(x, z)P(z, y)\\ &\leq \varliminf_{N\to\infty} \frac{1}{N} \sum_{n=0}^{N-1} \sum_{z\in E} P^n(x,z)P(z,y)\\ &= \lim_{N\to\infty} \frac{1}{N} \sum_{n=0}^{N-1} P^{n+1}(x,y)\\ &= P^*(x, y). \qquad (*)\end{aligned}$$

Let use denote the set of all states which are positive-recurrent with respect to P by R^+. (This set can be assumed to be non-empty, $P^* = 0$, if and only if $R^+ = \emptyset$ (cf. (i)). Taking into account the relation (*)

and the assertion (i) we obtain:

$$\sum_{y\in R^+} P^*(x, y) = \sum_{y\in E} P^*(x, y)$$

$$\geq \sum_{y\in E} \sum_{z\in E} P^*(x, z)P(z, y)$$

$$= \sum_{y\in E} \sum_{z\in R^+} P^*(x, z)P(z, y)$$

$$= \sum_{z\in R^+} \sum_{y\in E} P^*(x, z)P(z, y)$$

$$= \sum_{z\in R^+} P^*(x, z).$$

The interchange of the summation is possible because the double series absolutely converges. Hence, the relation $(P^*P)(x, y) = P^*(x, y)$ follows immediately for all $x, y \in E$.

<u>$PP^* = P^*$:</u> Let y be any positive-recurrent state (for example $y \in R_1$) and $x \in T$. Caused by

$$f^*(x, R_1) = \sum_{z\in R_1} P(x, z) + \sum_{z\in T} P(x,z)f^*(z, R_1)$$

it follows from lemma 1.1 and (iii) that

$$(PP^*)(x, y) = \sum_{z\in E} P(x, z)P^*(z, y)$$

$$= \sum_{z\in R_1} P(x, z)\pi(y) + \sum_{z\in T} P(x,z)f^*(z,R_1)\pi(y)$$

$$= f^*(x, R_1)\pi(y) = P^*(x, y).$$

For other combinations of states the proof of $(PP^*)(x, y) = P^*(x, y)$ becomes a simple consequence of lemma 1.1 and (iii).

<u>$(P^*)^2 = P^*$</u> can be proved analogously.

Obviously, the matrix P^* is substochastic. If P^* is stochastic, then P is called non-dissipative. If E is finite, then each transition matrix is non-dissipative. This is, for instance, a consequence of theorem 1.3 (iv).

Further, let us remark that the matrix P^*, which is defined as the component-wise CESARO-limit of the matrices P^n, always exists but, in general, it does not represent the CESARO-limit of the operators P^n exemplified by the matrix

$$P = \begin{pmatrix} 0 & 1 & 0 & 0 & \cdots \\ 0 & 0 & 1 & 0 & \cdots \\ 0 & 0 & 0 & 1 & \cdots \\ \cdot & \cdot & \cdot & \cdot & \\ \cdot & \cdot & \cdot & \cdot & \\ \cdot & \cdot & \cdot & \cdot & \end{pmatrix} .$$

For every $x \in E$ we obtain $\lim_{N\to\infty} \frac{1}{N} \sum_{n=0}^{N-1} (P^n \mathbf{1})(x) = 1$ but $P^* = 0$ and, consequently, $(P^* \mathbf{1})(x) = 0$. However, the following proposition is valid.

<u>Lemma 1.4.</u> If P is non-dissipative, then for any $f \in B(E)$ and $x \in E$ it holds that

$$\lim_{N\to\infty} \frac{1}{N} \sum_{n=0}^{N-1} (P^n f)(x) = (P^* f)(x).$$

<u>Proof:</u> Obviously, we obtain

$$\frac{1}{N} \sum_{n=0}^{N-1} (P^n f)(x) = \frac{1}{N} \sum_{n=0}^{N-1} \sum_{y\in E} P^n(x, y)f(y) = \sum_{y\in E} \frac{1}{N} \sum_{n=0}^{N-1} P^n(x, y)f(y).$$

Applying theorem 1.3 (iv) it follows that

$$\lim_{N\to\infty} \frac{1}{N} \sum_{n=0}^{N-1} (P^n f)(x) = \sum_{y\in E} (\lim_{N\to\infty} \frac{1}{N} \sum_{n=0}^{N-1} P^n(x, y))f(y) = \sum_{y\in E} P^*(x,y)f(y). \blacksquare$$

Now we shall direct our attention to the decomposition of transition matrices corresponding to the described decomposition of the state space. As above, let us denote the system of recurrence classes by $(R_i)_{i\in I}$ $(I \subset \mathbb{N})$ and the set of transient states by T with respect to the transition matrix P. According to the decomposition of the state space E, the matrix P can be decomposed into submatrices [1] $P_{R_iR_j}$, P_{R_iT}, P_{TR_j} and P_{TT} $(i, j \in I)$. Based on the properties of lemma 1.1 we obtain

$$P_{R_i} P_{R_j} = 0 \text{ [2]} \quad (i \neq j)$$

and

$$P_{R_iT} = 0$$

for all $i \in I$. Each of the submatrices $P_{R_iR_i}$ $(i \in I)$ is stochastic. In an analogous manner, any function $f \in B(E)$ can be described by its restrictions f_{R_i}, $i \in I$, and f_T.

<u>Lemma 1.5.</u>

(i) For all $n \in \mathbb{N}_0$ and $i \in I$ it holds that

---

[1]) When mentioning these submatrices the index sets are supposed to be non-empty by tacit agreement.

[2]) To simplify symbols we do not mark the size of null matrices.

$$(P^n)_{R_iR_i} = (P_{R_iR_i})^n,$$

$$(P^n)_{R_iR_j} = 0 \ (i \neq j),$$

$$(P^n)_{R_iT} = 0, \text{and}$$

$$(P^n)_{TT} = (P_{TT})^n .$$

(ii) Let any $f \in B(E)$ be given. Then it holds for all $i \in I$ that

$$Pf|_{R_i} = P_{R_iR_i} f_{R_i} ,$$

and for all $x \in T$:

$$(Pf|_T)(x) = (\sum_{i \in I} P_{TR_i} f_{R_i} + P_{TT} f_T)(x) \text{ [1]}. \qquad (1.2)$$

(iii) For all $i \in I$ there exists (in the sense of convergence by components)

$$(P^*)_{R_iR_i} = \lim_{N \to \infty} \frac{1}{N} \sum_{n=0}^{N-1} (P_{R_iR_i})^n.$$

(iv) Let Q be any transition matrix with $Q_{R_1E} = P_{R_1E}$. Then the submatrices satisfy the identity

$$(Q^*)_{R_1E} = (P^*)_{R_1E} .$$

In particular, R_1 forms a recurrent class with respect to Q.

(v) For all $n \in \mathbb{N}$ it holds that $(P-P^*)^n = P^n - P^*$.

<u>Proof:</u>
(i) There is nothing to prove for n = 0 and n = 1. Now let the proposition be valid for any $n \in \mathbb{N}$. Then we obtain that

$$(P^{n+1})_{R_iR_j} = \sum_{k \in I} (P^n)_{R_iR_k} P_{R_kR_j} + (P^n)_{R_iT} P_{TR_j} .$$

By induction it follows that

$$(P^{n+1})_{R_iR_j} = (P^n)_{R_iR_i} P_{R_iR_j} = (P_{R_iR_i})^n P_{R_iR_j} ,$$

i.e.,

---

[1]) One has to notice that, in general, the series (1.2) does not uniformly converge.

$$(P^{n+1})_{R_iR_j} = \begin{cases} (P_{R_iR_i})^{n+1} & i = j \\ 0 & i \neq j \end{cases} .$$

The other relations can be proved in an analogous manner.

(ii) These proposition can immediately be obtained by considering the structure of P supposed before.

(iii) This is a direct consequence of (i).

(iv) First we have to prove that

$$(Q^n)_{R_1E} = (P^n)_{R_1E} . \qquad (*)$$

There is nothing to prove for n = 0 and n = 1.
If the relation (*) is valid for any n, then we obtain:

$$\begin{aligned}(Q^{n+1})_{R_1R_j} &= \sum_{k\in I} (Q^n)_{R_1R_k} Q_{R_kR_j} + (Q^n)_{R_1T} Q_{TR_j} \\ &= \sum_{k\in I} (P^n)_{R_1R_k} Q_{R_kR_j} + (P^n)_{R_1T} Q_{TR_j} \\ &= (P^n)_{R_1R_1} Q_{R_1R_j} \\ &= (P^n)_{R_1R_1} P_{R_1R_j} = (P^{n+1})_{R_1R_j}\end{aligned}$$

for all $j \in I$. Similarly, it can be proved that

$$(Q^{n+1})_{R_1T} = (P^{n+1})_{R_1T} = 0.$$

Summarizing these results we obtain

$$(Q^*)_{R_1E} = (P^*)_{R_1E} .$$

Using the proposition (ii) of lemma 1.1 we can observe that every state recurrent with respect to P is also recurrent with respect to Q.

(v) We notice that the proposition does not need any proof for n = 1. Its validity for n + 1 follows by induction with respect to n using the equations (cf. theorem 1.3)

$$(P - P^*)^{n+1} = (P - P^*)^n(P - P^*)$$

and

$$\begin{aligned}(P^n - P^*)(P - P^*) &= P^{n+1} - P^nP^* - P^*P + (P^*)^2 \\ &= P^{n+1} - P^* . \blacksquare\end{aligned}$$

<u>Theorem 1.6.</u> Let $f \in B(E)$ be a function satisfying the equation $f = Pf$. If $T \neq E$, then the equation

$$f_{R_i} = c_i \mathbf{1}_{R_i}$$

is valid for every $i \in I$ with a suitable constant $c_i \in \mathbb{C}$.

Proof: It is known (see for example [11]) that a function $f' \in B(R_i)$ satisfying $P_{R_i R_i} f' = f'$ is a constant. Due to (i) of lemma 1.5 the assertion is true. ∎

Theorem 1.7. Let P be a non-dissipative transition matrix on E. Then it holds that:
(i) for all $N \in \mathbb{N}$ we have $\ker(I - P)^N = \ker(I - P^*)$,
(ii) $\dim(\ker(I - P)) < \infty$, if and only if the number of recurrence classes is finite. In this case the dimension of $\ker(I - P)$ corresponds with the number of recurrence classes.

Proof:
(i) Since the eigenvalue 1 of P has the index 1 (cf. appendix, lemma A 6) one obtains

$$\ker(I - P)^N = \ker(I - P)$$

for all $N \in \mathbb{N}$.
Now let $f \in \ker(I - P^*)$. This implies that

$$Pf = P(P^*f) = (PP^*)f = P^*f = f$$

due to lemmas A 4 and 1.3. Thus $f \in \ker(I - P)$. Conversely, if $f \in \ker(I - P)$ one obtains that $P^n f = f$ for all n by induction. Consequently, it holds that

$$\frac{1}{N} \sum_{n=0}^{N-1} P^n f = f$$

for all $N \in \mathbb{N}$. By lemma 1.4 we obtain $P^*f = f$. Hence $\ker(I - P) = \ker(I - P^*)$.

(ii) By the non-dissipativity assumption P^* is a stochastic matrix whose recurrence structure is equal to that of P, according to theorem 1.3. As above, let $(R_i)_{i \in I}$ be the system of recurrence classes and $T = E \setminus \bigcup_{i \in I} R_i$ the set of the transient states. Let f be any element of $\ker(I - P^*)$. Then, due to lemma 1.5 and theorem 1.6, we obtain $f_{R_i} = c_i \mathbf{1}_{R_i}$ with a suitable constant $c_i \in \mathbb{C}$ and

$$f_T = \sum_{i \in I} (P^*)_{TR_i} f_{R_i} + P^*_{TT} f_T = \sum_{i \in I} c_i (P^*)_{TR_i} \mathbf{1}_{R_i} .$$

Then by $f \to F(f) := (c_i)_{i \in I}$ a linear and bijective mapping $F: \ker(I - P^*) \to \mathbb{C}^I$ is given. ∎

In the following we shall consider some properties of transient states. For this purpose let P be a transition matrix on E with a non-empty set of transient states. Then we define a matrix $\Pi(x, y)$ by the mapping:

$$(x, y) \to \mathfrak{R}(x, y) := \sum_{n=0}^{\infty} (P_{TT})^n(x, y)$$

$(x, y) \in T \times T$. The finiteness of the values $\mathfrak{R}(x, y)$ is guaranteed (cf., e.g. [72], 5-3).

Definition 1.7. The submatrix P_{TT} of P is said to be absorbable if

$$\lim_{N\to\infty} ((P_{TT})^N \mathbf{1}_T)(x) = 0$$

for all $x \in T$ and 1-absorbable if

$$(\mathfrak{R} \mathbf{1}_T) = \sum_{n=0}^{\infty} ((P_{TT})^n \mathbf{1}_T)(x) < \infty$$

for all $x \in T$. P_{TT} is called strongly absorbable if P_{TT} is 1-absorbable and the function $\mathfrak{R} \mathbf{1}_T$ is bounded.

For an interpretation of the different properties of absorbability let us consider the function τ_R defined on A by

$$\tau_R(\omega) := \inf\{n \in \mathbb{N}: X_n(\omega) \in R\}.$$

This function describes the random time up to attaining the set R of recurrent states. (Without loss of generality the set $R = E \setminus T$ is assumed to be non-empty, otherwise P_{TT} cannot be absorbable.)

Lemma 1.8. For all $x \in T$ the following holds:

(i) $\mathbb{P}_x(\tau_R = N) = ((P_{TT})^{N-1} \mathbf{1}_T - (P_{TT})^N \mathbf{1}_T)(x)$, $N \in \mathbb{N}$,

(ii) $\mathbb{P}_x(\tau_R < \infty) = 1 - \lim_{N\to\infty} ((P_{TT})^N \mathbf{1}_T)(x)$,

and, for an absorbable P_{TT},

(iii) $\mathbb{E}_x \tau_R \leq \sum_{n=0}^{\infty} ((P_{TT})^n \mathbf{1}_T)(x)$.

Proof:

(i) Obviously, it holds that

$$\mathbb{P}_x(\tau_R = N) = ((P_{TT})^{N-1} P_{TR} \mathbf{1}_R)(x).$$

Since

$$((P_{TT})^{N-1} P_{TR} \mathbf{1}_R)(x) + ((P_{TT})^{N-1} P_{TT} \mathbf{1}_T)(x)$$
$$= ((P_{TT})^{N-1} P_{TE} \mathbf{1}_E)(x) = ((P_{TT})^{N-1} \mathbf{1}_T)(x)$$

we obtain

$$\mathbb{P}_x(\tau_R = N) = ((P_{TT})^{N-1} \mathbf{1}_T - (P_{TT})^N \mathbf{1}_T)(x).$$

(ii) From (i) the following results:

$$\mathbb{P}_x(\tau_R < \infty) = \lim_{N\to\infty} \sum_{n=1}^{N} \mathbb{P}_x(\tau_R = n)$$

$$= \lim_{N\to\infty} \sum_{n=1}^{N} ((P_{TT})^{n-1} \mathbf{1}_T - (P_{TT})^n \mathbf{1}_T)(x)$$

$$= \mathbf{1}_T(x) - \lim_{N\to\infty} ((P_{TT})^N \mathbf{1}_T)(x).$$

(iii) If P_{TT} is absorbable, then $P_x(\tau_R = \infty) = 0$.
Consequently, the expectation of τ_R is given by

$$\mathbb{E}_x \tau_R = \sum_{n=1}^{\infty} n \, \mathbb{P}_x(\tau_R = n).$$

Using the relations

$$\sum_{n=1}^{N} n \, \mathbb{P}_x(\tau_R = n) = \sum_{n=1}^{N} ((P_{TT})^{n-1} \mathbf{1}_T)(x) - N((P_{TT})^N \mathbf{1}_T)(x),$$

which follow from (i) by induction, we obtain

$$\sum_{n=1}^{\infty} n \, \mathbb{P}_x(\tau_R = n) \leq \sum_{n=1}^{N} ((P_{TT})^{n-1} \mathbf{1}_T)(x)$$

$$\leq \sum_{n=0}^{\infty} ((P_{TT})^n \mathbf{1}_T)(x). \blacksquare$$

Following lemma 1.8 we remark that the submatrix P_{TT} is absorbable, if and only if the Markov chain (X_n) stays in the set T of the transient states for only a finite time with probability 1, after having started in any state x. If P_{TT} is 1-absorbable, then the set T is already left up to a random time τ_R with a finite expectation. Moreover, this expectation $\mathbb{E}_x \tau_R$ is bounded with respect to $x \in T$ if P_{TT} is strongly absorbable.

Using the 1-absorbability we obtain the following result:

<u>Theorem 1.9.</u> If the submatrix P_{TT} is 1-absorbable, then it holds

$$\mathfrak{N}[(I - P_{TT})f] = f$$

for each function $f \in B(T)$.

<u>Proof:</u> Firstly, we remark that the expression on the left side is well-defined because $(I-P_{TT}) f$ is a bounded function. Since the matrix $|I - P_{TT}|$ is an element of $\mathfrak{R}(T)$ and f is bounded there exist the products $\mathfrak{N}(|I - P_{TT}||f|)$ and $(\mathfrak{N}|I - P_{TT}|)|f|$. Following the assertion 5-5 [72] it holds that $\mathfrak{N}(I - P_{TT}) = I$. Consequently, from lemma A 3 (appendix), the equation

$$\mathfrak{N}[(I - P_{TT})f] = [\mathfrak{N}(I - P_{TT})]f = f$$

results. $\blacksquare$

Definition 1.8. Let P be a non-dissipative transition matrix on E. P is said to be 1-recurrent if, in the case $T \neq \emptyset$, the submatrix P_{TT} of P is 1-absorbable. The transition matrix P is called a strong Markov chain if each class of recurrent states is strongly ergodic and, supposed $T \neq \emptyset$, the submatrix P_{TT} is absorbable.

Note: The notion of a "strong Markov chain" was introduced by ASO and KIMURA [1] and should not be confused with the strong Markov property.

Further, let us remark that by definition a 1-recurrent transition matrix is a non-dissipative one. However, the inversion is generally false as can be seen by the following

Example 1.1. Let $E = \mathbb{N}_0$ be the state space and let

$$P = \begin{pmatrix} 1 & 0 & 0 & 0 & \cdots \\ 1-a_1 & 0 & a_1 & 0 & \cdots \\ 1-a_2 & 0 & 0 & a_2 & \cdots \\ \vdots & \vdots & \vdots & \vdots & \end{pmatrix}$$

with

$$a_n := \frac{n}{n+1} \qquad (n = 1,2,\ldots).$$

Obviously, then

$$P^n(x, y) = \begin{cases} 1 - \prod_{k=x}^{x+n-1} a_k = 1 - \frac{x}{x+n} & \text{if } y = 0 \\ \prod_{k=x}^{n} a_k \qquad = \frac{x}{x+n} & \text{if } y = x + 1 + n \\ 0 & \text{otherwise} \end{cases}$$

Therefore, one obtains

$$\lim_{n\to\infty} P^n = P^* = \begin{pmatrix} 1 & 0 & 0 & \cdots \\ 1 & 0 & 0 & \cdots \\ 1 & 0 & 0 & \cdots \\ \vdots & \vdots & \vdots & \end{pmatrix}$$

(in the sense of component-wise convergence), this means that P is non-dissipative.

Since, by lemma 1.1, for all $x \in \mathbb{N}$ we have $\sum_{n=0}^{\infty} P^n(x,x) = 1$. Consequently, $T = \mathbb{N}$.

Now,

$$\Pi = \sum_{n=0}^{\infty} (P_{TT})^n = \begin{pmatrix} 1 & a_1 & a_1a_2 & a_1a_2a_3 & \cdots \\ 0 & 1 & a_2 & a_2a_3 & \cdots \\ 0 & 0 & 1 & a_3 & \cdots \\ 0 & 0 & 0 & 1 & \cdots \\ & & \cdots\cdots\cdots & & \end{pmatrix}$$

and we have

$$(\Pi \mathbf{1}_T)(1) = 1 + \sum_{n=1}^{\infty} \frac{1}{n+1} = \infty .$$

This implies that P_{TT} is not 1-absorbable.

The notation of 1-recurrence means an essential strengthening of that of non-dissipativity. The property of being a "strong Markov chain" is stronger than 1-recurrence, too. In the case of 1-recurrence it is possible that a recurrent state has the ergodic degree 1 and, consequently, the related recurrence class is not strongly ergodic. The existence of such a transition matrix was shown by KEMENY, SNELL and KNAPP [72]. Moreover, there are some examples of 1-absorbable submatrices which are not strongly absorbable. (cf. example 2.1 in chapter 2.)

1.2. QUASI-COMPACTNESS

Interpreting the transition matrix of a homogeneous Markov chain with the state space E as a bounded linear operator on B(E) (transition operator) we obtain some properties of the related Markov chain by the theory of linear operators. For each operator of this kind $\operatorname{spr} P = \|P\| = 1$ and $1 \in \sigma(P)$ obviously hold. Since $P\,\mathbf{1} = \mathbf{1}$, 1 is an eigenvalue with a multiplicity which is greater than or equal to the number of recurrence classes of P (cf. theorems 1.6 and 1.7).

In the present considerations a particular role is played by Markov chains whose transition operator is quasi-compact in the following sense:

Definition 1.9. An operator $T \in \mathfrak{L}(B)$ is said to be compact if, for each bounded sequence $(x_n)_{n\in\mathbb{N}}$ from B, the sequence $(Tx_n)_{n\in\mathbb{N}}$ has a convergent subsequence. An operator $T \in \mathfrak{L}(B)$ is said to be quasi-compact if there is a sequence $(K_n)_{n\in\mathbb{N}}$ of compact operators on $\mathfrak{L}(B)$ with the property

$$\lim_{n\to\infty} \|T^n - K_n\| = 0.$$

Compact and quasi-compact linear operators have many favourable properties. An operator induced by any transition matrix P is always compact if E is finite. The compactness, however, is a very stringent condition for infinite state spaces. Nevertheless, quasi-compactness can be guaranteed by the following condition:

Definition 1.10. A transition matrix P is said to fulfil the DOEBLIN condition if there exist a finite subset A of E, a positive integer N, and a real number $\alpha > 0$ such that

$$P^N(x, A) := \sum_{y \in A} P^N(x, y) \geq \alpha \qquad (D)$$

holds for all $x \in E$.

Theorem 1.10. [85] A transition matrix P fulfils the DOEBLIN-condition, if and only if the related linear operator $P \in \mathfrak{L}(B(E))$ is quasi-compact.

Thus, in the following a transition matrix is said to be quasi-compact if the related linear operator is quasi-compact. Such linear operators have the following spectral properties:

Lemma 1.11.

(i) A transition operator $P \in \mathfrak{L}(B(E))$ is quasi-compact, if and only if the set $\{\lambda \in \sigma(P): |\lambda| = 1\}$ consists of a finite number r of poles $\lambda_1, \ldots, \lambda_r$ each of them has a finite-dimensional eigenspace.

(ii) If a transition operator $P \in \mathfrak{L}(B(E))$ is quasi-compact, then each of its eigenvalues λ with $|\lambda| = 1$ is some root of unity.

Proof (cf. [123])

Theorem 1.12. Let P be a quasi-compact transition operator. Then

(i) $\lim_{N\to\infty} \frac{1}{N} \sum_{n=0} P^n = P^* = e_1(P)$ [1]

(the limit with respect to the $\mathfrak{L}(B(E))$-norm),

(ii) $\ker(I - P)^n = \ker(I - P^*)$ $(n = 1,2,\ldots)$,

(iii) $\sigma(P - P^*) = \sigma(P) \{1\}$.

Proof:

(i) By [123]

$$\lim_{N\to\infty} \frac{1}{N} \sum_{n=0}^{N-1} P^n = e_1(P).$$

---

[1] $e_1(P) := e_{\{1\}}(P)$, (cf. A 2, linear operators).

As it holds

$$\sum_{n=o}^{N-1} P^n \in \mathfrak{X}(\mathfrak{R}(E)),$$

$e_1(P) \in \mathfrak{X}(\mathfrak{R}(E))$ results by lemma A 11 (appendix). Following this proposition $\frac{1}{N}\sum_{n=o}^{N-1} P^n$ converges component-wise to a matrix which generates the operator $e_1(P)$. This matrix is, however, P^*.

(ii) Based on lemma A 6 (v) it holds that

$$\ker(I - P) = \ker(I - P)^n$$

for all $n \in \mathbb{N}$. From (i) it results that

$$\ker(I - P) = \ker(I - P^*).$$

(iii) is a direct consequence of (i) and of the spectral decomposition theorem A 8. ∎

Corollary 1.13. Let P be a quasi-compact transition operator. Then there exists the linear bounded operator

$$Z_\beta := (I - \beta(P - P^*))^{-1}$$

for all $\beta \in [0,1]$. The function $\beta \to Z_\beta$ is continuous for all $\beta \in (0,1]$.

Proof: In the case $\beta = 0$ there is nothing to prove. If $\beta \in (0,1]$, then $\beta^{-1} \in [1, \infty)$ and due to theorem 1.12 we have $\beta^{-1} \in \rho(P - P^*)$. Consequently, there exist

$$-\beta^{-1}((P - P^*) - \beta^{-1}I) = (\beta(\beta^{-1}I - (P - P^*)))^{-1}$$
$$= (I - \beta(P - P^*))^{-1} = Z_\beta.$$

Based on this representation and the fact that the resolvent

$$R(\beta^{-1}, P - P^*) = ((P - P^*) - \beta^{-1}I)^{-1}$$

is an analytic function, the continuity of the function $\beta \to Z_\beta$ follows.

Lemma 1.14. Let P be a quasi-compact transition operator. Then the following relations hold

(i) $ZP^* = P^*Z = P^*$ [1]),
and

(ii) $(I - P)Z = Z(I - P) = I - P^*$.

Proof: Due to lemma A 6 (iv) the series

$$-\sum_{n=o}^{\infty} \beta^{n+1}(P - P^*)^n$$

converges and is equal to $R(\beta^{-1}, P - P^*)$ for all $\beta \in (0, 1)$. Therefore,

---

[1]) For convenience in the following we set $Z := Z_1$.

we obtain

$$Z_\beta = \sum_{n=o}^{\infty} \beta^n (P - P^*)^n.$$

Since $(P - P^*)^n = P^n - P^*$, $n = 1,2,\ldots$, (cf. lemma 1.5 (v)) and $P^nP^* = P^*P^n = P^*$, $n = 1,2,\ldots$, (cf. lemma 1.3) it follows

$$Z_\beta = \sum_{n=o}^{\infty} \beta^n (P^n - P^*) + P^*$$

and

$$P^*Z_\beta = \sum_{n=o}^{\infty} \beta^n P^*(P^n - P^*) + (P^*)^2 = P^*.$$

Analogously, we find $Z_\beta P^* = P^*$.
According to lemma 1.13, the function $\beta \to Z_\beta$, $\beta \in (0,1]$, is continuous. Therefore, one obtains

$$P^*Z = P^* \lim_{\beta\uparrow 1} Z_\beta = \lim_{\beta\uparrow 1} P^*Z_\beta = P^*$$

and, in a similar manner, $ZP^* = P^*$.

Thus, (i) is proved. From this proposition the relations

$$\begin{aligned}(I - P)Z &= (I - (P - P^*))Z - P^*Z \\ &= Z^{-1}Z - P^*Z = I - P^*\end{aligned}$$

and, analogously, $Z(I - P) = I - P^*$ can be directly concluded.

<u>Lemma 1.15.</u> If P is a quasi-compact transition operator, then the matrix $I - (P - P^*)$ has a uniquely determined inverse

$$Z := (I - (P - P^*))^{-1} ,$$

which generates the operator $Z \in \mathfrak{L}(B(E))$ also referred to as Z.

<u>Proof:</u> For each $\beta \in [0, 1)$ it holds

$$Z_\beta = \lim_{N\to\infty} \sum_{n=o}^{\infty} \beta^n (P - P^*)^n \text{ and } Z = \lim_{\beta\uparrow 1} Z_\beta .$$

Obviously, each of the operators is an element of $\mathfrak{I}(\mathfrak{R}(E))$. As $\mathfrak{I}(\mathfrak{R}(E))$ is closed (cf. lemma A 11) all of the operators Z_β, $\beta \in [0, 1)$, and Z are also elements of $\mathfrak{I}(\mathfrak{R}(E))$. Now the assertion follows from lemma A11.∎

In the following we are interested in an explicit representation of the matrix Z, its associate operator, and its powers. For this purpose we need the following identities:

$$\frac{1}{(1-x)^N} = \sum_{n=o}^{\infty} \binom{n+N-1}{n} x^n \qquad (1.3)$$

($x \in [0, 1)$, $N \in \mathbb{N}$) and

$$\sum_{k=0}^{n} \binom{k+N-1}{k} = \binom{n+N}{n} \qquad (1.4)$$

($N, n \in \mathbb{N}_0$). The validity of these equations.is easily proven by a complete induction.

<u>Theorem 1.16.</u> Let P be a quasi-compact transition operator and let d be a positive integer, such that for every eigenvalue λ of P with $|\lambda| = 1$ it holds that $\lambda^d = 1$.

Then, for each $N \in \mathbb{N}$, it holds that

$$Z^n = \sum_{n=0}^{\infty} \binom{n+N-1}{n} \sum_{l=0}^{d-1} (P - P^*)^{nd+l} \qquad (1.5)$$

(with respect to $\mathfrak{L}(B(E))$), where the series (1.5) converges absolutely.

<u>Proof:</u> By theorem 1.11 there exists only a finite set of pairwise different numbers $\lambda_1, \dots, \lambda_q$ with $\lambda_i \in \sigma(P)$ and $|\lambda_i| = 1$ for $i = 1, \dots, q$. Without loss of generality, we assume $\lambda_1 = 1$. Each of the numbers λ_i is an isolated eigenvalue of P and, according to lemma 1.11, a root of the unity. Consequently, it is possible to find a number d which fulfils the conditions of the proposition.

We first prove the assertion for $N = 1$:
Let us denote $\sigma := \sigma(P) \{\lambda_1, \dots, \lambda_q\}$. According to the spectral decomposition theorem one obtains

$$P^n = \sum_{i=1}^{q} \lambda_i^n \, e\lambda_i(P) + P^n e_\sigma(P)$$

for $n \in \mathbb{N}$. With $\tilde{P} := Pe_\sigma(P)$ it obviously holds that $\tilde{P}^n = P^n e_\sigma(P)$ for all $n \in \mathbb{N}$.
Consequently, for each $n \in \mathbb{N}$, it follows that

$$\sum_{l=0}^{d-1} (P - P^*)^{nd+l} = \sum_{l=0}^{d-1} (P^{nd+l} - P^*) =$$

$$= \sum_{l=0}^{d-1} \Big(\sum_{i=1}^{q} \lambda_i^l \, e_{\lambda_i}(P) + \tilde{P}^{nd+l} - e_1(P) \Big) =$$

$$= \sum_{i=2}^{q} \sum_{l=0}^{d-1} \lambda_i^l \, e_{\lambda_i}(P) + \sum_{l=0}^{d-1} \tilde{P}^{nd+l}$$

by theorems 1.3 and 1.12. According to the special choice of d one obtains
$\sum_{l=0}^{d-1} \lambda_i^l = 0$ for $i = 2, \dots, q$ and, consequently,

$$\sum_{l=0}^{d-1} (P - P^*)^{nd+1} = \sum_{l=0}^{d-1} \tilde{P}^{nd+l} .$$

From the spectral decomposition theorem it results that

$$\sigma(\tilde{P}|_{e_\sigma(P)B(E)}) = \sigma$$

and, based on this equation, the relation spr $\tilde{P} < 1$ follows. Let $\gamma \in (\text{spr}\,\tilde{P}, 1)$ be an arbitrary number. By the definition of spr $\tilde{P}$ it is possible to find an $n_0 \in \mathbb{N}$ such that it holds for all $n \geq n_0$:

$$\|\tilde{P}^n\| \leq \gamma^n .$$

From this one obtains that

$$\|\sum_{l=0}^{d-1} \tilde{P}^{nd+l}\| \leq \sum_{l=0}^{d-1} \|\tilde{P}^{nd+l}\| \leq \sum_{l=0}^{d-1} \gamma^{nd+l}$$

and, for $m := [\frac{n_0}{d}]$ (the entire value),

$$\|\sum_{n=0}^{\infty} \sum_{l=0}^{d-1} (P - P^*)^{nd+l}\|$$

$$\leq \|\sum_{n=0}^{m} \sum_{l=0}^{d-1} (P - P^*)^{nd+l}\| + \sum_{n=m+1}^{\infty} \sum_{l=0}^{d-1} \gamma^{nd+l}$$

$$= \|\sum_{n=0}^{m} \sum_{l=0}^{d-1} (P - P^*)^{nd+l}\| + \frac{\gamma^{m+1}}{1-\gamma} < \infty .$$

Thus the absolute convergence of the series (1.5) with respect to $\mathfrak{L}(B(E))$ is proved. It remains to show that this series coincides with the operator Z. For all $n \in \mathbb{N}_0$ it holds that

$$(I -(P - P^*)) \sum_{l=0}^{d-1} (P - P^*)^{nd+l} = (P - P^*)^{nd} - (P - P^*)^{(n+1)d}$$

and

$$(I -(P - P^*)) \sum_{n=0}^{\infty} \sum_{l=0}^{d-1} (P - P^*)^{nd+l} = \sum_{n=0}^{\infty} (I -(P - P^*)) \sum_{l=0}^{d-1} (P - P^*)^{nd+l}$$

$$= \sum_{n=0}^{\infty} ((P - P^*)^{nd} - (P - P^*)^{(n+1)d}) = I.$$

In an analogous manner one obtains

$$(\sum_{n=0}^{\infty} \sum_{l=0}^{d-1} (P - P^*)^{nd+l})(I -(P - P^*)) = I.$$

Now, let us assume that the assertion is valid for any $N \in \mathbb{N}$. We set $T_n := \sum_{l=0}^{d-1} (P-P^*)^{nd+l}$ for $n \in \mathbb{N}_0$. As just shown it holds that

$$\|T_n\| \leq \sum_{l=0}^{d-1} \gamma^{nd+l} = \gamma^{nd} \, \frac{1 - \gamma^d}{1 - \gamma} .$$

By (1.3) it follows

$$\sum_{n=0}^{\infty} \binom{n+N'-1}{n} \|T_n\|$$

$$\leq \sum_{n=0}^{m} \binom{n+N'-1}{n} \|T_n\| + \frac{1-\gamma^d}{1-\gamma} \sum_{n=m+1}^{\infty} \binom{n+N'-1}{n} (\gamma^d)^n$$

$$\leq \sum_{n=0}^{m} \binom{n+N'-1}{n} \|T_n\| + \frac{1-\gamma^d}{1-\gamma} \sum_{n=0}^{\infty} \binom{n+N'-1}{n} (\gamma^d)^n$$

$$= \sum_{n=0}^{m} \binom{n+N'-1}{n} \|T_n\| + \frac{1}{(1-\gamma)(1-\gamma^d)^{N'-1}}$$

for all $N' \in \mathbb{N}$, i.e., the series $\sum_{n=0}^{\infty} \binom{n+N'-1}{n} T_n$ converges absolutely for each $N' \in \mathbb{N}$. In view of the assumption of induction, by (1.4), it results that

$$Z^{N+1} = Z^N Z = \left(\sum_{n=0}^{\infty} \binom{n+N-1}{n} T_n\right)\left(\sum_{k=0}^{\infty} T_k\right)$$

$$= \sum_{n=0}^{\infty} \left(\sum_{k=0}^{n} \binom{k+N-1}{k}\right) T_n = \sum_{n=0}^{\infty} \binom{n+N}{n} T_n.$$

Thus (1.5) is proved. ∎

<u>Remark 1.1.</u> As it can easily be seen from the proof of the representation (1.5) for $N = 1$, it holds that

$$\sum_{k=0}^{n} (P^k - P^*) = \sum_{k=0}^{n} \tilde{P}^k + \sum_{i=2}^{q} \sum_{j=0}^{s} \lambda_i^j \, e_{\lambda_i}(P)$$

for every $n \in \mathbb{N}$ with $s \equiv n \pmod d$ and $0 \leq s < d$. Since spr $\tilde{P} < 1$, it therefore follows that there is a number $B \in \mathbb{R}$ which satisfies the condition

$$\left\| \sum_{k=0}^{n} (P^k - P^*) \right\| \leq B \qquad (D^*)$$

for every $n \in \mathbb{N}_0$.

If the matrix P is quasi-compact, then the matrix $H := Z - P^*$ is well-defined and is associated with a bounded linear operator on $B(E)$.

<u>Theorem 1.17.</u> The matrix H has the following properties:

(i) $HP^* = P^*H = 0$,

(ii) $(I - P)H = H(I - P) = I - P^*$,

(iii) $H^N = Z^N - P^*$ $(N \in \mathbb{N})$,

(iv) for $f \in B(E)$ and any $N \in \mathbb{N}$ it holds that

$$H^N f|_{R_i} = (H^N)_{R_i R_i} f_{R_i} = (H_{R_i R_i})^N f_{R_i}$$

for each class R_i, $i \in I$, of recurrent states with respect to P,

(v) for each transient state $x \in E$, the relation

$$(H^N f)(x) = (H^N)_{TR(x)} f_{R(x)}(x) + (H^N)_{TT(x)} f_{T(x)}(x)$$

is valid with

$$R(x) := \{y \in E: x \rightsquigarrow y, \ y \text{ recurrent state}\}$$

$$T(x) := \{y \in E: x \rightsquigarrow y, \ y \text{ transient state}\},$$

(vi) $H_{TT} \geq 0$ and $H_{TT}(x,y) > 0$, if and only if $x \rightsquigarrow y$ holds.

Proof:

(i) From the assertion of lemma 1.14 it follows that

$$HP^* = (Z - P^*)P^* = ZP^* - P^* = 0$$

and, in an analogous manner, $P^* H = 0$.

(ii) By lemmas 1.3 and 1.4, it results that
$(I - P)H = (I - P)(Z - P^*) = (I - P)Z - (I - P)P^* = (I - P)Z = I - P^*$.
The second equation can be obtained in the same way.

(iii) In the case $N = 1$, the relation is obvious. If this assertion is true for any $N \in \mathbb{N}$, then these identities follow:
$H^{N+1} = H^N H = (Z^N - P^*)(Z - P^*) = Z^{N+1} - P^* Z - Z^N P^* + (P^*)^2$.
From lemmas 1.3 and 1.4 it results that $H^{N+1} = Z^{N+1} - P^*$.

(iv) According to theorem 1.16 and (iii) with a suitably chosen number $d \in \mathbb{N}$, it holds that

$$H^N = Z^N - P^* = \sum_{n=o}^{\infty} \binom{n+N-1}{n} \sum_{l=o}^{d-1} (P - P^*)^{nd+l} - P^*$$

$$= \sum_{n=o}^{\infty} \binom{n+N-1}{n} \sum_{l=o}^{\infty} (P^{nd+l} - P^*). \quad (*)$$

Thus we obtain

$$(H^N)_{R_i R_i} = \sum_{n=o}^{\infty} \binom{n+N-1}{n} \sum_{l=o}^{d-1} ((P^{nd+l})_{R_i R_i} - (P^*)_{R_i R_i})$$

by lemma 1.5 and, consequently,

$$(H^N)_{R_i R_i} = \sum_{n=o}^{\infty} \binom{n+N-1}{n} \sum_{l=o}^{d-1} (P^{nd+l}_{R_i R_i} - P^*_{R_i R_i})$$

$$= \sum_{n=o}^{\infty} \binom{n+N-1}{n} \sum_{l=o}^{d-1} (P_{R_i R_i} - P^*_{R_i R_i})^{nd+l} - P^*_{R_i R_i}. \quad (\ddagger)$$

If $N = 1$, then one obtains (cf. theorem 1.16):

$$H_{R_i R_i} = (I - (P_{R_i R_i} - P^*_{R_i R_i}))^{-1} - P^*_{R_i R_i} .$$

From this it results that

$$(H^N)_{R_iR_i} = (H_{R_iR_i})^N$$

by (iii) and (‡). Let f be any element of B(E). Then the relation (*) and the assertion of lemma 1.5 imply the relations

$$H^N f|_{R_i} = \sum_{n=o}^{\infty} \binom{n+N-1}{n} \sum_{l=o}^{d-1} (P^{nd+l} f|_{R_i} - P^* f|_{R_i})$$

$$= \sum_{n=o}^{\infty} \binom{n+N-1}{n} \sum_{l=o}^{d-1} ((P^{nd+l})_{R_iR_i} f_{R_i} - (P^*)_{R_iR_i} f_{R_i})$$

$$= (H^N)_{R_iR_i} f_{R_i} = (H_{R_iR_i})^N f_{R_i} .$$

The relation (v) also results from (*) by remarking that, for every state $y \in E$ with $x \not\to y$, the equation

$$P^{nd+l}(x,y) = P^*(x,y) = 0$$

holds.

(vi) As in (iv) one obtains

$$H_{TT} = \sum_{n=o}^{\infty} \sum_{l=o}^{d-1} ((P_{TT})^{nd+l} - (P_{TT})^*) = \sum_{n=o}^{\infty} \sum_{l=o}^{d-1} (P_{TT})^{nd+l} .$$

From this it follows that the assertion is obviously true. ∎

The following theorem will be used in the next chapter (section 2.5).

Theorem 1.18. Let P be a quasi-compact transition operator and let $L := P - I$. Then it holds that

$$R(\rho,L) = -\rho^{-1}P^* + \sum_{n=o}^{\infty} (-1)^{n+1} \rho^n H^{n+1}$$

for all $\rho \in \mathbb{C}$ with $0 < |\rho| \leq \|H\|^{-1}$.

Proof: Firstly, we remark that the equation

$$R(\rho,L) = (L - \rho I)^{-1} = ((P - I) - \rho I)^{-1} = (P - (1 + \rho)I)^{-1}$$
$$= R(1 + \rho, P)$$

implies the relation

$$\sigma(L) = \sigma(P) - 1 = \{\lambda \in \mathbb{C} : \lambda + 1 \in \sigma(P)\}.$$

In particular, zero is an isolated eigenvalue of L, because one is an isolated eigenvalue of P due to **lemma** 1.11. As was proved in [70] the resolvent function R(.,L) has the Laurent expansion

$$R(\rho,L) = -\frac{1}{\rho}\tilde{P} - \sum_{n=1}^{\infty} \rho^{-n-1} D^n + \sum_{n=0}^{\infty} \rho^n S^{n+1} \qquad (*)$$

in a pointed 0-neighbourhood, where by $\tilde{P}$ is the eigenprojection of L with respect to the eigenvalue zero, and S and D are bounded linear operators satisfying the relations

$$D = L\tilde{P},\ SL = LS = I - P^* \text{ and } S\tilde{P} = \tilde{P}S = 0. \qquad (\ddagger)$$

The convergence of the Laurent expansion (*) is guaranteed for all ρ with $0 < |\rho| < \|S\|^{-1}$.

Let Γ be a rectifiable Jordan curve containing $\sigma(L)$, and let f be an analytic function defined on a neighbourhood of the inner domain of Γ with

$$f(0) = 1 \text{ and } f|_{\sigma(L)\setminus\{0\}} = 0.$$

Then, by the definition of P and theorem 1.12, it follows that

$$\tilde{P} = -\frac{1}{2\pi i}\int_{\Gamma} f(\lambda)R(\lambda,L)d\lambda = -\frac{1}{2\pi i}\int_{\Gamma+1} f(\lambda-1)R(\lambda-1,L)d\lambda$$

$$= \frac{1}{2\pi i}\int_{\Gamma+1} f(\lambda-1)R(\lambda,P)d\lambda = e_1(P) = P^*.$$

Consequently, the relations ($\ddagger$) and lemma 1.3 imply $D = (P - I)P^* = 0$. Using lemma 1.13 we obtain that $(P - I)S = (P - P^* - I)S = I - P^*$. Based on the relations one has $S = -Z(I - P^*) = -H$.∎

The following two propositions will describe a connection between quasi-compact transition matrices and strong Markov chains.

<u>Lemma 1.19.</u> Let P be a non-reducible transition matrix and $n_0 \in \mathbb{N}$ with the property that all states $x \in E$ are recurrent with respect to P^{n_0} and have an ergodic degree, which is not less than two. Then the set E represents a strong ergodic recurrence class with respect to P.

<u>Proof:</u> Let us consider the canonical process (X_n) on $(A, \mathcal{A}) = (E^{\mathbb{N}_0}, \bigotimes_{\mathbb{N}_0} \mathfrak{P}(E))$. If q describes the vector of any initial distribution, then we denote the probability measures which are generated by q and P and P^{n_0}, respectively, on $\mathcal{A}$ by $\mathbb{P}_q$ and $\mathbb{P}_q^{n_0}$, respectively. Obviously, for all $n \in \mathbb{N}$ and any state x_0, it follows that

$$\mathbb{P}_q^{n_0}(X_n = x_0) = \mathbb{P}_q(X_{nn_0} = x_0).$$

Further, let $\tau'_{x_0}: A \to \mathbb{N} \cup \{\infty\}$ be the function

$$\tau'_{x_0}(\omega) := \inf(\{nn_0 : n \in \mathbb{N},\ X_{nn_0}(\omega) = x_0\} \cup \{\infty\}) \quad (\omega \in A).$$

Obviously, it holds that $\tau_{x_0} \leq \tau'_{x_0}$ and

$$\mathbb{P}^{n_0}_{x_0}(\tau_{x_0} = n) = \mathbb{P}_{x_0}(\tau'_{x_0} = nn_0)$$

($n \in \mathbb{N}$). As x_0 is a recurrent state with respect to P^{n_0} one obtains $\mathbb{P}^{n_0}_{x_0}(\tau_{x_0} < \infty) = 1$. By using the relations

$$\mathbb{P}_{x_0}(X_n = x_o \text{ for some } n \in \mathbb{N})$$
$$\geq \mathbb{P}_{x_0}(X_{nn_0} = x_0 \text{ for some } n \in \mathbb{N})$$
$$= \mathbb{P}^{n_0}_{x_0}(X_n = x_o \text{ for some } n \in \mathbb{N}) = 1$$

we find that x_0 is a recurrent state with respect to P. Thus we obtain

$$\mathbb{E}_{x_0} \tau^2_{x_0} \leq \mathbb{E}_{x_0}(\tau'_{x_0})^2 = \sum_{n=1}^{\infty} (nn_0)^2 \, \mathbb{P}_{x_0}(\tau'_{x_0} = nn_0)$$
$$= n_o^2 \sum_{n=1}^{\infty} n^2 \, \mathbb{P}^{n_0}_{x_0}(\tau_{x_0} = n) = n_o^2 \, \mathbb{E}^{n_0}_{x_0} \tau^2_{x_0} < \infty ,$$

that means x_0 has an ergodic degree of at least order two. From lemma 1.2 and the assumed non-reducibility of P the assertion follows.∎

Theorem 1.20. Each quasi-compact transition matrix is a strong Markov chain (cf. definition 1.8) with a finite number of recurrence classes.

Proof: Let P be a quasi-compact transition matrix. Then, by theorem 1.10, there exist a finite set $A \subset E$, an integer $n_0 \in N$, and a number $\alpha > 0$ with $P^{n_0}(x, A) \geq \alpha$ for all $x \in E$. HORDIJK [55] proved that P is non-dissipative. Hence, P has at least one class of recurrent states, but at most a finite number of such classes by (D) (definition 1.10). Let R be such a class of recurrent states and let $A' := R \cap A$. From lemma 1.5 and (D), it results that $P^{n_0}_{RR}(x, A') \geq \alpha$ for all $x \in A$. That means $\tilde{P} := P_{RR}$ is quasi-compact on B(R). According to lemma 1.11 and the spectral mapping theorem (cf. [28]), $\tilde{P}^d$ is a quasi-compact matrix with $\{\lambda \in \sigma(\tilde{P}^d) : |\lambda| = 1\} = \{1\}$ for a suitable integer $d \in \mathbb{N}$.

Now it is easy to see that there are only recurrent states with respect to $\tilde{P}^d$. Furthermore, by theorem 1.16 and lemma 1.15, there exists the matrix

$$Z_d := \sum_{n=o}^{\infty} (\tilde{P}^d - (\tilde{P}^d)^*)^n = (I - (\tilde{P}^d - (\tilde{P}^d)^*))^{-1} .$$

Based on the proposition 9-75 [72], every recurrence class with respect to $\tilde{P}^d$ is strongly ergodic. Hence, by lemma 1.19, so is R. If the set of transient states with respect to P is non-empty, then it results $\operatorname{spr} P_{TT} < 1$ (cf. corollary 2.13 [123]). Thus the series

$$\mathfrak{N} = \sum_{n=o}^{\infty} (P_{TT})^n$$

converges in $\mathfrak{L}(B(T))$, and $\mathfrak{N} \, \mathbb{1}_T \in B(T)$ holds, i.e., P_{TT} is strongly absorbable.∎

1.3. CLUSTERING

Now let us consider two countable sets E and E' together with a surjective mapping $\varphi: E \to E'$ (i.e. card $E \geq$ card E'). As above, let (X_n) be the canonical process on $(A, \mathfrak{A}) = (E^{\mathbb{N}_0}, \bigotimes_{\mathbb{N}_0} \mathfrak{P}(E))$. Furthermore, we introduce the process (Y_n) on $(A, \mathfrak{A})$ given by $Y_n = \varphi(X_n)$, $n = 0, 1, 2, \ldots$

In this connection φ is called a cluster mapping or a lumpation of the state E, and (Y_n) is said to be **generated** by clustering or lumpation of (X_n). In short, (Y_n) is called a cluster process or image process and $\varphi(x) \in E'$ ($x \in E$) a cluster state.

For any fixed transition matrix P on E and an initial distribution $q \in \mathfrak{M}^1(E)$ the process (X_n) is a **homogeneous** Markov chain with respect to the measure $\mathbb{P}_q$ generated by q and P. On the other hand, the process (Y_n) needs, in general, not to be Markovian, as can easily be seen by some examples.

Definition 1.11. An $E \times E$-transition matrix P is said to be strongly lumpable with respect to a function $g: E \to E'$ if, for each initial distribution $q \in \mathfrak{M}^1(E)$, the process $(Y_n) = (g(X_n))$ on $(A, \mathfrak{A}, \mathbb{P}_q)$ is a homogeneous Markov chain with a transition matrix independent of q.

If P is strongly lumpable, then we shall say that "P allows a strong lumpation". The question arising, whether P is strongly lumpable may be answered by the following proposition. For this purpose let us denote:

$$\varphi^{-1}(s) := \{x \in E: g(x) = s\}$$

($s \in E'$) and, as above, for any $A \subset E$ and all $x \in E$

$$P(x,A) := \sum_{y \in A} P(x,y).$$

Theorem 1.21.

(i) P is strongly lumpable, if and only if

$$P(x, \varphi^{-1}(s)) = P(y, \varphi^{-1}(s)), \quad (s \in E), \tag{1.6}$$

holds for all $x,y \in E$ with $\varphi(x) = \varphi(y)$.

(ii) If an $E \times E$-transition matrix allows a strong lumpation, then an $E' \times E'$-transition matrix P is defined by the relation

$$\overline{P}(s,t) := P(x, g^{-1}(t))$$

for any $x \in \varphi^{-1}(s)$ and $s,t \in E'$. For each initial distribution $q \in \mathfrak{M}^1(E)$ the matrix $\overline{P}$ is the transition matrix of the homogeneous Markov chain $(Y_n) = (\varphi(X_n))$ on $(A, \mathfrak{A}, \mathbb{P}_q)$.

Note: In the case of a finite state space this proposition was proved by KEMENY and SNELL [71]. The sufficiency of the condition (1.6) for the strong lumpability was shown by DYNKIN [29] and ZESSIN [126] in the case of a more general state space.

Proof (theorem 1.21):

(i) Necessity of (1.6): Let x,y be states with $\varphi(x) = \varphi(y) =: s$, then

$$\mathbb{P}_x(Y_0 = s) = \mathbb{P}_y(Y_0 = s) = 1.$$

Since P is strongly lumpable the equation

$$\mathbb{P}_x(Y_1 = t | Y_0 = s) = \mathbb{P}_y(Y_1 = s | Y_0 = s)$$

has to be true for all $t \in E'$. However, it holds that

$$\mathbb{P}_x(Y_1 = t | Y_0 = s) = \frac{\mathbb{P}_x(Y_0=s,\ Y_1=t)}{\mathbb{P}_x(Y_0=s)} = P(x, \varphi^{-1}(t))$$

and, analogously,

$$\mathbb{P}_y(Y_1 = t | Y_0 = s) = \frac{\mathbb{P}_y(Y_0=s,\ Y_1=t)}{\mathbb{P}_y(Y_0=s)} = P(y, \varphi^{-1}(t)),$$

implying (1.6).

(ii) Sufficiency of (1.6): If (1.6) is fulfilled, then the $E' \times E'$-matrix P is well-defined, because $\bar{P}(s,t)$ is not dependent on the special choice of the element $x \in \varphi^{-1}(s)$. Obviously, we have

$$\sum_{t \in E'} \bar{P}(s,t) = \sum_{t \in E'} \sum_{z \in \varphi^{-1}(t)} P(x,z) = \sum_{z \in E} P(x,z) = 1$$

and $\bar{P}(s,t) \geq 0$ for any $x \in \varphi^{-1}(s)$ and for all $s,t \in E'$.

In this way, a transition matrix is given by $\bar{P} = (\bar{P}(s,t))_{s,t \in E'}$. Now let $q \in \mathfrak{M}^1(E)$ be an initial distribution on E. Let (φq) be the initial distribution on E' defined by

$$(\varphi q)(s) := q(\varphi^{-1}(s)).$$

Then, for all $n \geq 2$ and $s_0, \ldots, s_n \in E'$,

$$\mathbb{P}_q(Y_0 = s_0, \ldots, Y_n = s_n) = \sum_{x_0 \in \varphi^{-1}(s_0)} \cdots \sum_{x_n \in \varphi^{-1}(s_n)} q(x_0)P(x_0,x_1) \ldots \times$$

$$\times P(x_{n-1}, x_n)$$

$$= \sum_{x_0 \in \varphi^{-1}(s_0)} \cdots \sum_{x_n \in \varphi^{-1}(s_n)} q(x_0)P(x_0,x_1) \ldots \times$$

$$\times P(x_{n-2}, x_{n-1})\bar{P}(s_{n-1}, s_n)$$

$$= \mathbb{P}_q(Y_0=s_0, \ldots, Y_{n-1} = s_{n-1})\overline{P}(s_{n-1}, s_n)$$

by (1.6). Consequently, we obtain – if $\mathbb{P}_q(Y_0=s_0,\ldots,Y_{n-1} = s_{n-1}) \neq 0$ – the relation

$$\mathbb{P}_q(Y_n=s_n \mid Y_{n-1} = s_{n-1}, \ldots, Y_0 = s_0) = \overline{P}(s_{n-1}, s_n).$$

In the same way, one obtains

$$\mathbb{P}_q(Y_1 = s_1, Y_0 = s_0) = q(s_0)\overline{P}(s_0, s_1),$$

i.e.,

$$\mathbb{P}_q(Y_1 = s_1 \mid Y_0 = s_0) = \overline{P}(s_0, s_1)$$

if $\mathbb{P}_q(Y_0 = s_0) \neq 0$. Thus (Y_n) is a homogeneous Markov chain with the transition matrix P independent of q. Therefore, (ii) is valid, too. ∎

The condition (1.6) is, of course, very restrictive. Therefore, let us remark that also other notions of lumpability which require weaker conditions can be given: A transition matrix P is called **completely** lumpable with respect to φ if (Y_n) is a **(non-necessarily homogeneous)** Markov chain on $(A, \mathfrak{A}, \mathbb{P}_q)$ for every choice of an initial distribution $q \in \mathfrak{M}^1(E)$. P is said to be weakly lumpable with respect to φ and q if (Y_n) is a Markov chain on $(A, \mathfrak{A}, \mathbb{P}_q)$ for the fixed initial distribution q . From the papers of ROSENBLATT [93], KEMENY and SNELL [71], and DIETZ and NOLLAU [25] it can be seen that each of these definitions, in fact, differs from **each other. There are the following implications:**

"P is strongly lumpable" $\Rightarrow$ "P is completely lumpable" $\Rightarrow$ "P is weakly lumpable",

where none of them can be reversed. In particular, examples of a completely lumpable transition matrix P can be given such that the transition **behaviour** of (Y_n) properly depends on q and is "purely inhomogeneous", respectively (cf. [94]). Here we are particularly interested in the notion of **strong** lumpability since it is very suitable for the considerations in the following.

Subsequently, by $\mathfrak{C} := \varphi^{-1}(\mathfrak{E}') = \varphi^{-1}(\mathfrak{P}(E'))$ let us denote the sub-σ-algebra of $\mathfrak{E}$ generated by φ. Further, let $\Phi f := f \circ \varphi$ for every $f \in B(E')$.

<u>Lemma 1.22.</u> Φ is a norm-preserving isomorphism of the Banach algebra $B(E', \mathfrak{E}')$ on the Banach algebra $B(E, \mathfrak{C})$ with the property that $\Phi\, \mathbf{1} = \mathbf{1}$.

<u>Proof:</u> Let f be any element of $B(E', \mathfrak{E}')$. Then we have

$$(\Phi f)^{-1}(\mathfrak{B}_{\mathbb{C}}) = \varphi^{-1}(f^{-1}(\mathfrak{B}_{\mathbb{C}})) \subset \varphi^{-1}(\mathfrak{E}') = \mathfrak{C}, \quad ^{1)}$$

---

1) $\mathfrak{B}_{\mathbb{C}}$ – system of all Borel sets $B \subset \mathbb{C}$.

that means Φf is a $\mathfrak{C}$-measurable function on E. Based on surjectivity of φ, it holds that

$$\sup_{x\in E} |(\Phi f)(x)| = \sup_{x\in E} |f(\varphi(x))| = \sup_{y\in E'} |f(y)| = \|f\| \qquad (*)$$

and, consequently, $\Phi f \in B(E, \mathfrak{C})$. Obviously, the mapping is linear and multiplicative. By (1.6), it also holds that $\|\Phi f\| = \|f\|$ for all $f \in B(E', \mathfrak{C}')$. The identity $\Phi \mathbf{1} = \mathbf{1}$ is trivial. Hence, it remains to show that Φ is bijective.

Let $g \in B(E, \mathfrak{C})$. From lemma 55.1 [4] it follows that there is an $(\mathfrak{C}', \mathfrak{B}_{\mathbb{C}})$-measurable function $g_\varphi : E' \to \mathbb{C}$ with $g = g_\varphi \circ \varphi$. The function g_φ is uniquely determined on $\varphi(E) = E'$. In view of:

$$\sup_{y\in E'} |g_\varphi(y)| = \sup_{x\in E} |g_\varphi(\varphi(x))| = \|g_\varphi \circ \varphi\| = \|g\|$$

we remark that g_φ is an element of $B(E', \mathfrak{C}')$. For each $g \in B(E, \mathfrak{C})$ there exists a uniquely determined function $g_\varphi \in B(E', \mathfrak{C}')$ with $g = \Phi g_\varphi$. Thus we obtain the bijectivity of Φ.

Note: Obviously, the operator Φ is monotone in the following sense: For each real function $f \in B(E')$,

$$f \geq 0 \ (> 0) \iff \Phi f \geq 0 \ (> 0).$$

Let L be any operator of $\mathfrak{L}(B(E'))$. Then by $\tilde{\Phi}L$ we denote the following operator on $B(E, \mathfrak{C})$:

$$(\tilde{\Phi}L)f := \Phi L \Phi^{-1} f \quad (f \in B(E, \mathfrak{C})).$$

Lemma 1.23. $\tilde{\Phi}$ defines a norm-preserving isomorphism of the Banach algebra $\mathfrak{L}(B(E'))$ onto the Banach algebra $\mathfrak{L}(B(E, \mathfrak{C}))$ with $\tilde{\Phi}I = I$.

Before proving this lemma we shall introduce the following notations. The identity operator will be denoted by I independently of the considered space. Moreover, we shall use the following abbrevations:

$$\mathfrak{L} := \mathfrak{L}(B(E)), \ \mathfrak{L}' := \mathfrak{L}(B(E', \mathfrak{C}')), \text{ and } \mathfrak{L}_{\mathfrak{C}} := \mathfrak{L}(B(E, \mathfrak{C})).$$

Proof (lemma 1.23): Let $L \in \mathfrak{L}'$. Then the operator $\tilde{\Phi}L = \Phi L \Phi^{-1}$, composed of the bounded linear operators

$$\Phi^{-1} : B(E, \mathfrak{C}) \to B(E'),$$
$$L : B(E') \to B(E'), \text{ and}$$
$$\Phi : B(E') \to B(E, \mathfrak{C})$$

(cf. lemma 1.22), is itself a bounded linear mapping of $B(E, \mathfrak{C})$ into $(B, \mathfrak{C})$. This means $\tilde{\Phi}$ is a mapping $\mathfrak{L}' \to \mathfrak{L}_{\mathfrak{C}}$. With the aid of lemma 1.22 (guaranteeing that both the operators are norm-preserving) we see

$$\|\tilde{\Phi}L\| = \sup_{\substack{\|f\|\leq 1 \\ f\in B(E,\mathbb{C})}} \|\tilde{\Phi}Lf\| = \sup_{\substack{\|f\|\leq 1 \\ f\in B(E,\mathbb{C})}} \|\Phi L\Phi^{-1}f\|$$

$$= \sup_{\substack{\|f\|\leq 1 \\ f\in B(E,\mathbb{C})}} \|L\Phi^{-1}f\| = \sup_{\substack{\|g\|\leq 1 \\ g\in B(E')}} \|Lg\| = \|L\| .$$

These relations show that $\tilde{\Phi}$ is norm-preserving, too. $\tilde{\Phi}$ is an isomorphism, because

$$\begin{aligned}\tilde{\Phi}(a_1 L_1 + a_2 L_2)(f) &= \Phi((a_1L_1 + a_2L_2)\Phi^{-1}f)\\ &= \Phi(a_1L_1 \Phi^{-1}f + a_2L_2 \Phi^{-1}f)\\ &= a_1 \Phi L_1 \Phi^{-1}f + a_2 \Phi L_2 \Phi^{-1}f\\ &= a_1(\tilde{\Phi}L_1)(f) + a_2(\tilde{\Phi}L_2)(f)\end{aligned}$$

holds for all $f \in B(E, \mathbb{C})$, $a_1,a_2 \in \mathbb{C}$ and $L_1,L_2 \in \mathfrak{L}'$. Moreover, we obtain

$$\tilde{\Phi}(SL) = \Phi SL \Phi^{-1} = \Phi S\Phi^{-1} \Phi L \Phi^{-1} = \Phi(\tilde{\Phi}S)(\tilde{\Phi}L).$$

This means that $\tilde{\Phi}$ is a linear and multiplicative operator with $\tilde{\Phi}I = \Phi I \Phi^{-1} = \Phi\Phi^{-1} = I$.
The relation $\tilde{\Phi}S = \tilde{\Phi}L$ $(S,L \in \mathfrak{L}')$ is equivalent to $\Phi S \Phi^{-1}f = \Phi L \Phi^{-1}f$ (for all $f \in B \in B(E,\mathbb{C})$) and $\Phi Sg = \Phi Lg$ for all $g \in B(E')$, respectively, because Φ^{-1} is surjective. Since Φ is injective, it follows $Sg = Lg$ for all $g \in B(E')$, establishing the injectivity of $\tilde{\Phi}$.

The surjectivity of $\tilde{\Phi}$ will be obtained as follows:
For any $L \in \mathfrak{L}_{\mathbb{C}}$ the operator $\tilde{L} := \Phi^{-1}L\Phi$ is linear and bounded on $B(E')$, because Φ^{-1}, L and Φ have these properties. Then it holds that

$$\tilde{\Phi}\tilde{L} = \Phi(\Phi^{-1}L\Phi)\Phi^{-1} = L. \blacksquare$$

Lemma 1.24. An $E \times E$-transition matrix P is strongly lumpable if and only if the inclusion

$$PB(E,\mathbb{C}) \subset B(E,\mathbb{C})$$

holds.

Proof: The condition (1.6) is equivalent to the following one: For all $s \in E'$,

$$P\, \mathbf{1}_{\varphi^{-1}(s)} \in B(E,\mathbb{C}). \qquad (*)$$

If $PB(E,\mathbb{C}) \subset B(E,\mathbb{C})$, then this condition is automatically fulfilled, and, therefore, P is strongly lumpable by theorem 1.21.

Conversely, the relation (*) implies the inclusion $PB(E,\mathbb{C}) \subset B(E,\mathbb{C})$ as the following considerations will show. From (*) the relation $Pf \in B(E,\mathbb{C})$ immediately results for each finite linear combination f of indicator functions $\mathbf{1}_{\varphi^{-1}(s)}$. (Such a function f will be called a simple function). If $f \in B(E,\mathfrak{C})$ and $f \geq 0$, then there exists a certain sequence (f_n) of non-negative simple functions with $f = \sup_{n\in\mathbb{N}} f_n$. By the theorem of the monotone convergence (see for example [4]) one obtains that

$$Pf = P(\sup_{n\in\mathbb{N}} f_n) = \sup_{n\in\mathbb{N}} Pf_n ;$$

hence, Pf is $\mathfrak{C}$-measurable. For any real-valued function $f \in B(E,\mathfrak{C})$ the representation $Pf = Pf^+ - Pf^-$ implies that $Pf \in B(E,\mathfrak{C})$, because $Pf^+ \in B(E,\mathfrak{C})$ and $Pf^- \in B(E,\mathfrak{C})$. Finally, the assertion of the lemma will be obtained by the decomposition of an arbitrary function into its real and imaginary part, respectively.∎

Using the notation $P_{\mathfrak{C}}$ for the restriction of P on $B(E,\mathfrak{C})$ we can reformulate lemma 1.23 as follows:

<u>Corollary 1.25.</u> P allows a strong lumpation, if and only if $P_{\mathfrak{C}} \in \mathfrak{L}_{\mathfrak{C}}$.

The connection between the isomorphism considered in the lemmas 1.22 and 1.23 will be illustrated by the following

<u>Theorem 1.26.</u> Let the $E \times E$-transition matrix P be strongly lumpable. Then it holds that

$$\widetilde{\Phi}\bar{P} = P_{\mathfrak{C}} ,$$

where $\bar{P}$ defined in theorem 1.21 is the $E' \times E'$-transition matrix taken as an operator on $B(E', \mathfrak{C}')$.

<u>Proof:</u> Let $g \in B(E')$. Then the following relations hold for all $x \in E$:

$$\begin{aligned}(\widetilde{\Phi}\bar{P})(\Phi g)(x) &= (\Phi\bar{P}\,\Phi^{-1})(\Phi g)(x) = (\Phi\bar{P})(x)\\ &= \sum_{t\in E'} \bar{P}(\varphi(x),t)g(t)\\ &= \sum_{t\in E'} \sum_{y\in\varphi^{-1}(t)} P(x,y)g(\varphi(y))\\ &= \sum_{y\in E} P(x,y)g(\varphi(y)) = (P\Phi g)(x).\end{aligned}$$

It follows that $(\widetilde{\Phi}\bar{P})f = Pf$ for all $f \in B(E,\mathfrak{C})$ and $\widetilde{\Phi}\bar{P} = P_{\mathfrak{C}}$, respectively.∎

<u>Lemma 1.27.</u> Let P be a strongly lumpable $E \times E$-transition matrix. Then every power P^n of P, $n \in \mathbb{N}$, allows a strong lumpation with

$$\bar{P}^n = (\overline{P^n}) \text{ and } (P_{\mathfrak{C}})^n = (P^n)_{\mathfrak{C}} .$$

<u>Proof:</u> $PB(E,\mathfrak{C}) \subset B(E,\mathfrak{C})$ by lemma 1.24, because P is strongly lumpable. Then, the relation $P^nB(E,\mathfrak{C}) \subset B(E,\mathfrak{C})$ follows by induction. That means P^n is strongly lumpable, too. Now the identity $(P_{\mathfrak{C}})^n = (P^n)_{\mathfrak{C}}$ is also simply verifiable by induction. Hence, from the proposition of theorem 1.26, it results

$$\bar{P}^n = \Phi^{-1}(P_{\mathfrak{C}})^n = \Phi^{-1}(P_n)_{\mathfrak{C}} = (\overline{P^n})$$

for all $n \in \mathbb{N}$. ∎

Let us now turn to the question to what extent properties of a strongly lumpable transition matrix P are preserved with respect to the lumpated matrix P. Firstly, quasi-compactness of the associated operators are of particular interest.

Lemma 1.28. Let P be a strongly lumpable $E \times E$-transition matrix. Then it holds that:
(i) The operator $\bar{P}$ is quasi-compact, if and only if the operator $P_{\mathfrak{C}}$ is quasi-compact.
(ii) If P is quasi-compact, then $\bar{P}$ and $P_{\mathfrak{C}}$ are quasi-compact.

Proof:
(i) Let $\lambda \in \rho(\bar{P})$. Then there exists the linear bounded operator $(\bar{P} - \lambda I)^{-1} = R(\lambda, \bar{P})$. Based on theorem 1.26 and the identity

$$\tilde{\Phi}(\bar{P} - \lambda I) = P_{\mathfrak{C}} - \lambda I,$$

the equations

$$I = \tilde{\Phi}((\bar{P} - \lambda I)R(\lambda, \bar{P})) = \tilde{\Phi}(\bar{P} - \lambda I)\tilde{\Phi}R(\lambda, \bar{P})$$
$$= (P_{\mathfrak{C}} - \lambda I)\tilde{\Phi}R(\lambda, \bar{P})$$

follow, and one obtains analogously:

$$I = \tilde{\Phi}R(\lambda, \bar{P})(P_{\mathfrak{C}} - \lambda I).$$

Hence, we have $\lambda \in \rho(P_{\mathfrak{C}})$ and $\tilde{\Phi}R(\lambda, \bar{P}) = R(\lambda, P_{\mathfrak{C}})$. Likewise the relation $\rho(P_{\mathfrak{C}}) \subset \rho(\bar{P})$ follows (by consideration of $\tilde{\Phi}^{-1}$ instead of $\tilde{\Phi}$). Thus one obtains $\sigma(\bar{P}) = \sigma(P_{\mathfrak{C}})$. In view of theorem 1.12, the proposition is proved.
(ii) P satisfies the DOEBLIN condition by theorem 1.10. I.e., there are a finite set $A \subset E$, an integer $N \in \mathbb{N}$, and a number $\alpha > 0$ with

$$P^N(x, A) \geq \alpha$$

for all $x \in E$. Obviously, the set $A' := \{s \in E' : \varphi^{-1}(s) \cap A \neq \emptyset\}$ is finite, and lemma 1.26 implies:

$$\bar{P}^N(s, A') = \overline{P^N}(s, A') = \sum_{z \in \varphi^{-1}(A')} P^N(x, z)$$
$$\geq \sum_{z \in A} P^N(x, z) \geq \alpha.$$

I.e., $\bar{P}$ satisfies the DOEBLIN condition and, consequently, is quasi-compact. ∎

Notes:

1. The second proposition of lemma 1.28 does not allow an inversion. Every $E \times E$-matrix is strongly lumpable with respect to the (constant) mapping $\varphi: E \to \{1\}$. For instance, the transition matrix of the (deterministic) one-sided walk:

$$P = \begin{pmatrix} 0 & 1 & 0 & 0 & \cdots \\ 0 & 0 & 1 & 0 & \cdots \\ 0 & 0 & 0 & 1 & \cdots \\ \cdot & \cdot & \cdot & \cdot & \\ \cdot & \cdot & \cdot & \cdot & \\ \cdot & \cdot & \cdot & \cdot & \end{pmatrix}$$

is a counter-example. Each state is transient. P is not quasi-compact, because it holds that $\sigma(P) = \{\lambda \in \mathbb{C}: |\lambda| \leq 1\}$. On the other hand, the matrix $\bar{P} = (1)$ is a compact one.

2. We have to remark that the assertions 1.26 - 1.28 are formulated for the operator $P_{\mathfrak{C}}$ but not for an associated matrix. Although the operator $P_{\mathfrak{C}}$ can always be described by a matrix (e.g., by P), the matrix describing $P_{\mathfrak{C}}$ is not necessarily defined uniquely. If P is strongly lumpable and P is quasi-compact, then the operators

$$\bar{P}^* := \lim_{N\to\infty} \frac{1}{N} \sum_{n=0}^{N-1} \bar{P}^n$$

(in the sense of the norm of $\mathfrak{L}'$),

$$\bar{Z} := (I - (\bar{P} - \bar{P}^*))^{-1},$$

and

$$\bar{H} := \bar{Z} - \bar{P}^*$$

are well-defined, and it holds that $\bar{P}^*, \bar{Z}, \bar{H} \in \mathfrak{L}'$.

Theorem 1.29. Let P be a strongly lumpable and quasi-compact $E \times E$-transition matrix.

(i) Then $B(E,\mathfrak{C})$ is an invariant subspace of $B(E,\mathfrak{E})$ with respect to each of the operators P^*, Z and H.

(ii) $\tilde{\Phi}\bar{P}^* = (P^*)_{\mathfrak{C}}$, $\tilde{\Phi}\bar{Z} = Z_{\mathfrak{C}}$ and $\tilde{\Phi}\bar{H} = H_{\mathfrak{C}}$.

(iii) $(P^*)_{\mathfrak{C}} = (P_{\mathfrak{C}})^* := \lim_{N\to\infty} \frac{1}{N} \sum_{n=0}^{N-1} (P_{\mathfrak{C}})^n.$

(iv) The relation $(H_{\mathfrak{C}})^n = (H^n)_{\mathfrak{C}}$ holds for all $n \in \mathbb{N}_0$.

Proof:

(i) Let $f \in B(E,\mathfrak{C})$. Then it holds that

$$P^* f = \lim_{N\to\infty} \frac{1}{N} \sum_{n=0}^{N} P^n f.$$

Due to lemma 1.27 it results that $P^n f \in B(E,\mathfrak{C})$ for all $n \in \mathbb{N}$. As a limit of $\mathfrak{C}$-measurable functions P^*f is $\mathfrak{C}$-measurable, too, i.e., $P^*f \in B(E,\mathfrak{C})$. With a suitably chosen integer $d \in \mathbb{N}$ one obtains that

$$Zf = \sum_{n=0}^{\infty} \sum_{l=0}^{d-1} (P - P^*)^{nd+l} f,$$

because P is quasi-compact as assumed. Obviously, then it holds that $(P - P^*)^{nd+l} \in B(E,\mathfrak{C})$ for each $n \in \mathbb{N}$ and for all $l = 0,\dots,d-1$. Hence $Zf \in B(E,\mathfrak{C})$, too. The proposition on H results from $H = Z - P^*$.

(ii) The continuity of the isomorphism $\tilde{\Phi}$ was proved by lemma 1.23. This and the propositions of theorem 1.26 and lemma 1.27 imply

$$\tilde{\Phi}\bar{P}^* = \tilde{\Phi}(\lim_{N\to\infty} \frac{1}{N} \sum_{n=0}^{N-1} P^n) = \lim_{N\to\infty} \frac{1}{N} \sum_{n=0}^{N-1} \tilde{\Phi}\bar{P}^n$$

$$= \lim_{N\to\infty} \frac{1}{N} \sum_{n=0}^{N-1} (P_n)_{\mathfrak{C}} = (P^*)_{\mathfrak{C}} .$$

According to the theorem 1.16, the operators $\bar{Z}$ and Z have the following representations:

$$\bar{Z} = \sum_{n=0}^{\infty} \sum_{l=0}^{d-1} (\bar{P} - \bar{P}^*)^{nd+l}$$

and

$$Z = \sum_{n=0}^{\infty} \sum_{l=0}^{d-1} (P - P^*)^{nd+l},$$

respectively, with a suitably chosen integer $d \in \mathbb{N}$. Using the continuity of $\tilde{\Phi}$ again one obtains

$$\tilde{\Phi}\bar{Z} = \sum_{n=0}^{\infty} \sum_{l=0}^{d-1} \tilde{\Phi}(\bar{P} - \bar{P}^*)^{nd+l} ,$$

and the additivity and multiplicity allow the following representation

$$\tilde{\Phi}\bar{Z} = \sum_{n=0}^{\infty} \sum_{l=0}^{d-1} (\tilde{\Phi}\bar{P} - \tilde{\Phi}\bar{P}^*)^{nd+l} = \sum_{n=0}^{\infty} \sum_{l=0}^{d-1} ((P - P^*)_{\mathfrak{C}})^{nd+l}.$$

Moreover, $((P - P^*)_{\mathfrak{C}})^{nd+l} = ((P - P^*)^{nd+l})_{\mathfrak{C}}$ follows by induction, because $B(E,\mathfrak{C})$ is invariant with respect to $P - P^*$.
Thus we have

$$\tilde{\Phi}\bar{Z} = (\sum_{n=0}^{\infty} \sum_{l=0}^{d-1} (P - P^*)^{nd+l})_{\mathfrak{C}} = Z_{\mathfrak{C}} .$$

Finally, the relation $H = Z - P^*$ immediately implies that

$$\tilde{\Phi}\bar{H} = \tilde{\Phi}(\bar{Z} - \bar{P}^*) = Z_{\mathfrak{C}} - (P^*)_{\mathfrak{C}} = H_{\mathfrak{C}} .$$

(iii) From lemma 1.27 we obtain:

$$\frac{1}{N} \sum_{n=0}^{N-1} (P_{\mathfrak{C}})^n = (\frac{1}{N} \sum_{n=0}^{N-1} P^n)_{\mathfrak{C}} ,$$

and it follows that

$$(P_{\mathbb{C}})^* = \lim_{N\to\infty} \frac{1}{N} \sum_{n=0}^{N-1} (P_{\mathbb{C}})^n = (\lim_{N\to\infty} \frac{1}{N} \sum_{n=0}^{N-1} P^n)_{\mathbb{C}} = (P^*)_{\mathbb{C}} \; . \blacksquare$$

In the following we give some statements concerning the classification of states with respect to the lumpated matrix $\bar{P}$.

Lemma 1.30. Let P be a strongly lumpable $E \times E$-transition matrix and $s,t \in E'$. Then the following implications are valid:

(i) $s \curvearrowright t \Rightarrow \forall x \in \varphi^{-1}(s) \; \exists y \in \varphi^{-1}(t): x \curvearrowright y$,

(ii) $\exists x \in \varphi^{-1}(s) \; \exists y \in \varphi^{-1}(t): x \curvearrowright y \Rightarrow s \curvearrowright t$.

Proof:

(i) If $s \curvearrowright t$, then $\bar{P}^{n_0}(s,t) > 0$ for a certain integer $n_0 \in \mathbb{N}_0$. Taking into account lemma 1.27, $P^{n_0}(x, \varphi^{-1}(t)) > 0$. Hence there is a state $y \in \varphi^{-1}(t)$ such that $P^{n_0}(x,y) > 0$, i.e., $x \curvearrowright y$.

(ii) If $x \curvearrowright y$, then $P^{n_0}(x,y) > 0$ for some integer $n_0 \in \mathbb{N}_0$. This implies:

$$\bar{P}^{n_0}(s,t) = P^{n_0}(x, \varphi^{-1}(t)) \geq P^{n_0}(x,y) > 0,$$

that is $s \curvearrowright t$. ∎

Lemma 1.31. Let P be a strongly lumpable $E \times E$-transition matrix.

(i) If P is non-dissipative, then $\bar{P}$ is non-dissipative, too, and it holds that $\overline{P^*} = \bar{P}^*$.

(ii) A cluster state $s \in E'$ is recurrent (with respect to $\bar{P}$) if there exists a recurrent state $x \in \varphi^{-1}(s)$ (with respect to P).

(iii) If P is non-dissipative, and $s \in E'$ is a recurrent cluster state (with respect to $\bar{P}$), then there exists a recurrent state $y \in \varphi^{-1}(s)$ (with respect to P) with $x \curvearrowright y$ for every $x \in \varphi^{-1}(s)$.

(iv) If P is non-dissipative, then a cluster state $s \in E'$ is transient (with respect to $\bar{P}$), if and only if all states $x \in \varphi^{-1}(s)$ are transient (with respect to P).

Proof:

(i) If P is non-dissipative, then a simple modification of theorem 1.3 (iv) shows that the series

$$\sum_{y\in\varphi^{-1}(t)} \frac{1}{N} \sum_{n=0}^{N-1} P^n(x,y)$$

is uniformly convergent with respect to $N \in \mathbb{N}$ for each $x \in E$ and $t \in E'$. Therefore, due to lemma 1.27, it follows:

$$\bar{P}^*(s,t) = \lim_{N\to\infty} \frac{1}{N} \sum_{n=o}^{N-1} \overline{P^n}(s,t)$$

$$= \lim_{N\to\infty} \sum_{y\in\varphi^{-1}(t)} \frac{1}{N} \sum_{n=o}^{N-1} P^n(x,y)$$

$$= \sum_{y\in\varphi^{-1}(t)} \lim_{N\to\infty} \frac{1}{N} \sum_{n=o}^{N-1} P^n(x,y)$$

$$= \sum_{y\in\varphi^{-1}(t)} P^*(x,y),$$

i.e. $\bar{P}^* = \overline{P^*}$. In particular, we have

$$\sum_{t\in E'} \bar{P}^*(s,t) = \sum_{t\in E'} \sum_{y\in\varphi^{-1}(t)} P^*(x,y) = \sum_{y\in E} P^*(x,y) = 1.$$

Thus $\bar{P}$ is non-dissipative.

(ii) Let $s \in E'$ and $x \in \varphi^{-1}(s)$ a recurrent state (with respect to P). Then, due to the lemmas 1.27 and 1.1 (ii), we obtain the relations:

$$\sum_{n=o}^{\infty} \bar{P}^n(s,s) = \sum_{n=o}^{\infty} \overline{P^n}(s,s) = \sum_{n=o}^{\infty} \sum_{y\in\varphi^{-1}(s)} P^n(x,y)$$

$$\geq \sum_{n=o}^{\infty} P^n(x,y) = \infty.$$

Consequently, the cluster state s is a recurrent one (with respect to $\bar{P}$) by lemma 1.1 (ii).

(iii) If P is non-dissipative, then part (i) shows the non-dissipativity of $\bar{P}$. Now let $s \in E'$ be a recurrent cluster state. This implies $\bar{P}^*(s,s) > 0$ by theorem 1.3. Then according to (i) it holds that

$$\bar{P}^*(s,s) = \sum_{y\in\varphi^{-1}(s)} P^*(x,y) > 0$$

for every state $x \in \varphi^{-1}(s)$. Hence, there exists a state $y \in \varphi^{-1}(s)$ with $P^*(x,y) > 0$. This state y is recurrent (with respect to P) due to theorem 1.3, and it holds that $x \frown y$.

(iv) This assertion results from (ii) and (iii) in the following way: If $x \in E'$ is a transient cluster state, then each state $x \in \varphi^{-1}(s)$ has to be transient by (ii). On the other hand, if $\varphi^{-1}(s)$ contains only transient states, then each $y \in \varphi^{-1}(y)$ with $x \frown y$ is transient for any $x \in \varphi^{-1}(s)$. By (iii), this is only possible if s is transient. ∎

2. MARKOV DECISION PROBLEMS (MDP)

2.1. THE MODEL

Definitions: Let

(i) a non-empty, at most a countable set $\mathbb{E}$,

(ii) a non-empty set $\mathfrak{D}$, and

(iii) a function $D: \mathbb{E} \to \mathfrak{P}(\mathfrak{D})$, satisfying the following condition: for every $x \in \mathbb{E}$ the set $D(x)$ is non-empty and finite,
be given.
With $K := \{(x,d): x \in \mathbb{E}, d \in D(x)\}$, let be given

(iv) a function $p: K \times \mathbb{E} \to [0, 1]$ having the property

$$\sum_{y \in \mathbb{E}} p(x,d,y) = 1$$

for every $(x,d) \in K$ and

(v) a function $r: K \to \mathbb{R}$.

Definition 2.1. The tuple

$$\mathfrak{M} := (\mathbb{E}, \mathfrak{D}, D, p, r)$$

is called a (stationary) Markov Decision Model (MDM).

Firstly we give the following interpretation of an MDM. Let a system randomly moving within a set $\mathbb{E}$ of possible states be observed at the moments $t = 0,1,2,\ldots$ $\mathbb{E}$ is called the state space of the system. If a state $x \in \mathbb{E}$ is observed at the moment t, then a decision d can be chosen from the set $D(x)$ of decisions admissible in the state x. As a consequence of the choice of d the system moves into a "new" state up to moment $t+1$ according to the probability distribution defined by $p(x,d,.)$. Simultaneously, the observer receives (or expects) a reward $r(x,d)$.

Due to this interpretation we call $\mathfrak{D}$ a "decision space", $D(x)$ the set of decisions admissible in the state x, p the "transition law" of the system, and r the "reward structure".

In this paper we require the following general assumption:
Let there exist a number $k \in [0, \infty)$, satisfying

$$0 \leqq r(x,d) \leqq k$$

for all $(x,d) \in K$.

Now we introduce the notation of a policy. For every $n \in \mathbb{N}_0$, let $\mathbb{A}_n := \mathbb{E}^{n+1}$. $\mathbb{A}_n$ can be understood as the set of possible "histories" of the system up to the moment n.

Definition 2.2. A policy is a sequence $(f_0, f_1, \ldots, f_n, \ldots)$ of functions $f_n: \mathbb{A}_n \to \mathfrak{D}$ with the property

$$f_n(x_0,\ldots,x_n) \in D(x_n)$$

for every $n \in \mathbb{N}_0$, $(x_0,\ldots,x_n) \in \mathbb{A}_n$.

Hence, a policy is a rule giving the decision to be made at every moment with respect to all observations of the system up to this moment. The set of all policies is denoted by $\mathfrak{R}$.

Subsequently, $(\mathbb{A}, \mathcal{A}) = (E^{\mathbb{N}_0}, \bigotimes_{\mathbb{N}_0} \mathfrak{P}(E))$ denotes the canonical basic space and $(X_n)_{n\in\mathbb{N}_0}$ the family of canonical projections defined on $(\mathbb{A}, \mathcal{A})$ (see 1.1). It is possible to endow the space $(\mathbb{A}, \mathcal{A})$ with a probability measure in dependence on any given "initial distribution" $\mu \in \mathfrak{M}^1(E)$ and policy $\pi \in \mathfrak{R}$. This will be done here in the following way:
For every cylinder set $[X_0 = x_0, \ldots, X_n = x_n]$, $n \in \mathbb{N}_0$, $x_0, \ldots, x_n \in E$, let

$$\mathbb{P}^{\pi}_{\mu}(X_0 = x_0, \ldots, X_n = x_n) :=$$

$$= \mu(x_0)p(x_0, f_0(x_0), x_1) \cdots p(x_{n-1}, f_{n-1}(x_0, \ldots, x_{n-1}), x_n).$$

Since the σ-algebra $\mathcal{A}$ is generated by the system of these cylinder sets it is possible to extend the so-defined function $\mathbb{P}^{\pi}_{\mu}$ to a probability measure on $\mathcal{A}$. The existence of this measure follows from a well-known extension theorem. Another well-known theorem can show this measure to be uniquely defined (cf. [4]). For simplicity, we denote this probability measure by $\mathbb{P}^{\pi}_{\mu}$, too. (Analogously to the construction used in 1.1 this construction described above is, to a certain extent, a non-Markovian one.)
The process $(X_n)_{n\in\mathbb{N}_0}$ is a stochastic process on each of the probability spaces $(\mathbb{A}, \mathcal{A}, \mathbb{P}^{\pi}_{\mu})$, $\mu \in \mathfrak{M}^1(E)$, $\pi \in \mathfrak{R}$, whose finite-dimensional distributions, as a rule, properly depend on the policy applied. Therefore, $(X_n)_{n\in\mathbb{N}_0}$ is called a "Markov Decision Process". If the policy $\pi \in \mathfrak{R}$ is applied, we call $(X_n)_{n\in\mathbb{N}_0}$ on $(\mathbb{A}, \mathcal{A}, \mathbb{P}^{\pi}_{\mu})$ the "process ruled by π". For the rest of this chapter we write (X_n) for $(X_n)_{n\in\mathbb{N}_0}$. Subsequently the index μ of $\mathbb{P}^{\pi}_{\mu}$ will be omitted if it is clear (or without significance) which initial distribution is under consideration.

Notes:

1. The notion of a Markov decision process is due to BELLMAN [5]. In the literature the related term "Markovian sequential control process" is used as well ([20]).

2. Regardless of the term "Markov Decision Process", which is connected with the structure of the transition law, the process (X_n) on $(\mathbb{A}, \mathcal{A}, \mathbb{P}^{\pi})$ is, in general, no Markovian process.

3. The MDM defined here can be considered as a special case of the "stochastic dynamic decision model" introduced by HINDERER [49], which admits the transition law and the reward structure as well as the sets of admissible decisions to be dependent on the whole history of the system. With regard to the optimality criteria which are treated in this paper and to which the state space transformation will be applied, we confine ourselves to the stationary Markov case.

4. In several papers, e.g., of BLACKWELL [8], DERMAN [20], [18], [19], ROSS [96], [97], MAITRA [81], GUBENKO and STATLAND [39], IWAMOTO [63] and HORDIJK [54], optimality considerations are not based on the policy set $\mathfrak{R}$, but on the more extensive class of so-called "randomized policies", however, the assumptions required there and the finiteness of the decision sets $D(x)$, $x \in \mathbb{E}$, guarantee the optimum to be reached on $\mathfrak{R}$ in each case.

The following definitions introduce classes of policies with special properties. Here the term of a decision function plays a central role.

Definition 2.3. A function $f: \mathbb{E} \to \mathfrak{D}$ satisfying $f(x) \in D(x)$ for every $x \in \mathbb{E}$ is called a decision function. The set of all decision functions is denoted by $\mathbb{D}$.

Definition 2.4. A policy $\pi = (f_0', f_1', \ldots, f_n', \ldots)$ is called a Markov policy if there is a sequence $(f_n)_{n \in \mathbb{N}_0}$ of decision functions for which the equations

$$f_n'(x_0, \ldots, x_n) = f_n(x_n)$$

are valid for every $n \in \mathbb{N}_0$, $(x_0, \ldots, x_n) \in \mathbb{A}_n$. The set of all Markov policies is denoted by $\mathfrak{R}_M$.

Definition 2.5. A policy $\pi = (f_0', f_1', \ldots, f_n', \ldots)$ is called stationary if there is a decision function satisfying

$$f_n'(x_0, \ldots, x_n) = f(x_n)$$

for every $n \in \mathbb{N}_0$, $(x_0, \ldots, x_n) \in \mathbb{A}_n$. We denote the set of all stationary policies by $\mathfrak{R}_S$.

It is obvious that every stationary policy is also a Markov policy. Subsequently, for the sake of brevity, every Markov policy π is identified with the corresponding sequence of decision functions unless **explicitly** stated otherwise in the context. Therefore, a stationary policy is also written in the form

$$\pi = (f, f, \ldots, f, \ldots) =: f^{\infty} \quad (f \in \mathbb{D}).$$

Since the definitions given above are independent of p and r, we now introduce

Definition 2.6. The triple $(\mathbb{E}, \mathfrak{D}, D)$ is called a decision field.

As it is easy to see, the process (X_n) turns out to be a Markov process on $(\mathbb{A}, \mathfrak{A}, \mathbb{P}^{\pi})$ if it is ruled by a Markov policy $\pi \in \mathfrak{R}_M$. If π is stationary, then (X_n) even becomes a homogeneous Markov process (which, in general, does not need to be stationary). If, for every decision function $f \in \mathbb{D}$, we denote by

$$P(f) := (p(x, f(x), y))_{x, y \in \mathbb{E}}$$

the transition matrix induced by f (the elements of which will also be written as $P(f)(x,y)$, $x,y \in E$) it is easy to see that the transition of the process ruled by $\pi = (f_0,f_1,\dots,f_n,\dots)$ from the moment n to the moment n + 1 is described by the matrix $P(f_n)$. Therefore, the n-step transition probabilities of this process can be collected in the following matrices:

$$Q_0(\pi) := I, \text{ and}$$

$$Q_n(\pi) := Q_{n-1}(\pi)P(f_{n-1}) \quad (n \in \mathbb{N}).$$

Add to this, hereafter the notion of a "reward function" on E, induced by a decision function $f \in D$ according to

$$x \to r(f)(x) := r(x,f(x)), \; x \in E,$$

will frequently be used. Because of the assumptions required for the reward structure r, every reward function $r(f)$, $f \in \mathbb{D}$, is non-negative and bounded. In the following the latter - $r(f) \in B(E)$ - will be of essential significance.

REWARD FUNCTIONS AND NOTIONS OF OPTIMALITY

Our first aim in considering Markov Decision Processes is to compute those policies which ensure a "maximal gain" (with respect to certain criteria) connected with the sequence of observations and corresponding decisions. The second aim consists in computing this "maximal gain". The notion of "gain" has to be defined precisely in accordance to the problem to be solved. There are several possibilities to do this in connection with a suitable notion of an optimal policy; various of them were thoroughly investigated in the literature.

In the present paper we shall consider four of the most important notions of optimality.

1. N-step-optimality

If we want to maximize the gain obtained during the first N steps (with a fixed $N \in \mathbb{N}_0$) of the Markov Decision Process, we proceed as follows: Every initial state $x \in E$ is connected with the expectation of the pay-off sum, which is obtained "in mean" by applying the policy π until the time N after the process started in x.

Definition 2.7. Let $\pi \in \mathcal{R}$ and let $N \in \mathbb{N}_0$ be arbitrary. The function

$$x \to V_N^\pi(x) := \mathbb{E}_x^\pi \sum_{n=0}^{N} r(X_n,f_n(X_0,\dots,X_n)) \;{}^{1)}, \; x \in E,$$

is called the N-step (expected) reward of the policy π. The number N is called "planning horizon".

---

1) $\mathbb{E}_x^\pi$ denotes the expectation with respect to the measure $\mathbb{P}_{\varepsilon_x}^\pi$.

Note: V_N^π does exist because r is bounded. Moreover, the functions V_N^π are uniformly bounded in $\pi \in \mathfrak{R}$.

Definition 2.8. $V_N^* := \sup_{\pi \in \mathfrak{R}} V_N^\pi$ is called optimal N-step (expected) reward.

It is reasonable that a policy is called optimal if its N-step reward is optimal.

Definition 2.9. Let $C \subset \mathfrak{R}$. A policy $\pi^* \in C$ is called N-step optimal (on C) if $V_N^{\pi^*} \geq V_N^\pi$ for all $\pi \in C$. (The additional term "on C" will be omitted if $C = \mathfrak{R}$.)

In the section 2.2 we shall give statements concerning the existence and the problem of how to find N-step-optimal policies.

2. Discounted optimality

In some practical problems the influence of a discount factor β, $\beta \in [0,1)$, on the future pay-offs has to be taken into account. This leads to

Definition 2.10. The function

$$x \to V_\beta^\pi(x) := E_x^\pi \sum_{n=0}^{\infty} \beta^n r(X_n, f_n(X_0, \ldots, X_n)), \; x \in E,$$

is called the β-discounted (expected) total reward (in short: discounted total reward) of the policy $\pi \in \mathfrak{R}$.

Note: Boundedness of r implies uniform boundedness of V_β^π with respect to $\pi \in \mathfrak{R}$. More precisely, for $\beta \in [0,1)$ and $\pi \in \mathfrak{R}$ it holds that $V_\beta^\pi \in B(E)$, and

$$\|V_\beta^\pi\| \leq (1-\beta)^{-1} k.$$

Definition 2.11. $V_\beta^* := \sup_{\pi \in \mathfrak{R}} V_\beta^\pi$ is called optimal β-discounted (expected) total reward.

Now we can introduce the

Definition 2.12. Let $C \subset \mathfrak{R}$ and $\beta \in [0,1)$. A policy $\pi^* \in C$ is called β-discounted optimal (on C) if $V_\beta^{\pi^*} \geq V_\beta^\pi$ for all $\pi \in C$.

Again, we omit "on C" if $C = \mathfrak{R}$. For brevity, the term "β-optimal" will often be used instead of "β-discounted optimal". Section 2.2 is concerned with the existence and computation of such policies.

3. Average optimality

If the planning horizon of an undiscounted problem is very large, it seems to be useful to assume it to be infinite. In this case the comparison of policies by means of the corresponding expected total rewards is impossible since, in general, the latter need not be infinite.

To avoid this difficulty we can introduce a discount factor and refer to the discounted problem as an approximation of the N-step problem with N being infinite (cf. HINDERER [50]). With fixed β this approximation has the disadvantage that the influence of policies becomes irrelevant after a sufficiently long time. Therefore, it is reasonable to choose another quite different approach. It essentially consists in quantifying not the total expected reward itself, but its "increase rate". This can be done by a CESARO-limitation procedure, where the planning horizon tends to infinite as follows.

<u>Definition 2.13.</u> For an arbitrary policy $\pi \in \mathfrak{R}$ the function

$$x \to g^{\pi}(x) := \liminf_{N \to \infty} \frac{1}{N} \sum_{n=0}^{N-1} \mathbb{E}_x^{\pi}\, r(X_n, f_n(X_o, \dots, X_n))$$

($x \in E$) is called average (expected) reward of π.

Note that the functions g^{π}, $\pi \in \mathfrak{R}$, do exist and are uniformly bounded in $\pi \in \mathfrak{R}$ by the boundedness assumption with respect to r.

<u>Definition 2.14.</u> $g^* := \sup_{\pi \in \mathfrak{R}} g^{\pi}$ is called optimal average (expected) reward.

<u>Definition 2.15.</u> Let $C \subset \mathfrak{R}$. A policy $\pi^* \in C$ is said to be average optimal (on C) if $g^{\pi^*} \geqq g^{\pi}$ for all $\pi \in C$.

Average optimality will be considered in the sections 2.3 and 2.4.

<u>4. N-discounted (sensitive) optimality</u>

It is easy to see that, in contrast to the situation in the discounted case, the influence of an average optimal policy may be advantageous only after a "very long time". In addition, examples can be given. They show that the average reward criterion is not very sensitive and that there might be some preferable policies in class of average optimal policies (cf. VEINOTT [116]). To avoid this disadvantage two classes of "sensitive" optimality criteria were introduced: the class of "sensitive averaging" optimality criteria and the class of "sensitive discount" optimality criteria. The former is based on CESARO-limitation procedures of higher order as in the case of the average (expected) reward and were treated, e.g., by VEINOTT [116], SLADKY [105], LIPPMAN [76], FLYNN [34], DENARDO [14], [15], and HORDIJK and SLADKY [58]. The latter which will be considered in this book are concerned with the N-th order increase rate of the β-discount (expected) reward if the discount factor β tends to one.

<u>Definition 2.16.</u> Let $C \subset \mathfrak{R}$ and $N \in \mathbb{N}_{-1}$. A policy $\pi^* \in C$ is called N-discount optimal on C if for every $\pi \in C$

$$\lim_{\beta \to 1} \inf\left(\frac{1-\beta}{\beta}\right)^{-N} (V_\beta^{\pi^*} - V_\beta^\pi) \geq 0.$$

Note: "N-discount optimal" is also called "N-optimal" or "sensitive optimal" in the sequel. Under certain assumptions (-1)-optimality is equivalent to average optimality, and 0-optimality coincides with "bias optimality" (cf. DENARDO [15], [14]) and "average overtaking optimality" (cf. LIPPMAN [76], FLYNN [34]), respectively. Subsequently, the functions V_N^π, V_β^π, and g^π are also called reward functions. If $\pi \in \Re$ is a stationary policy of the form $\pi = f^\infty$, $f \in \mathbb{D}$, we write V_N^f, V_β^f, and g^f instead of V_N^π, V_β^π, and g^π, respectively. If $\pi = f^\infty$ is an optimal policy in some sense, then we call f an optimal decision function.
Finally, let us note that there are more optimality notions which have been investigated in numerous papers. Some of them will be mentioned here without going into details: BLACKWELL [9] introduced the notation of B-optimality (cf. MAITRA [82]) which can be considered as a strengthening of sensitive discount optimality. JAQUETTE [64], [65], MANDL [83], and QUELLE [91] dealt with the so-called moment optimality.

Following the relevant literature, we call the field of problems concerning the optimization of policies with respect to the optimality criteria introduced above a "Markov Decision Problem" (MDP).

2.2. OPTIMAL POLICIES FOR DISCOUNTED AND FINITE HORIZON MDP's

In the fundamental papers of BELLMAN [6] and HOWARD [60], a system of recursive equations was developed for the N-step optimal rewards V_N^*, $N \in \mathbb{N}_0$. This is of basic importance in the Markov Decision Theory. These and the related equations concerning other optimality criteria are known als BELLMAN equations. By means of them both the optimal reward functions and the corresponding N-step optimal policies can be computed:

Theorem 2.1. [75] It holds that

$$V_0^* = \max_{f \in \mathbb{D}} r(f) \qquad (2.1)$$

and

$$V_N^* = \max_{f \in \mathbb{D}} \{r(f) + P(f)V_{N-1}^*\} \qquad (2.2)$$

for $N \in \mathbb{N}$.

Theorem 2.2. [75] Let

$$M_0 := \{f^* \in \mathbb{D}: r(f^*) = V_0^*\} \text{ and}$$
$$M_n := \{f^* \in \mathbb{D}: r(f^*) + P(f^*)V_{n-1}^* = V_n^*\}$$

for $N \in \mathbb{N}_0$, and let $N \in \mathbb{N}_0$ be arbitrary and $f_0^*, \ldots, f_N^* \in \mathbb{D}$ with $f_i^* \in M_i$, $i = 0, \ldots, N$.

Then every policy π^* of the form

$$\pi^* = (f_N^*, f_{N-1}^*, \ldots, f_1^*, f_o^*, f_{N+1}, f_{N+2}, \ldots)$$

with arbitrary decision functions $f_{N+1}, f_{N+2}, \ldots$ is N-step optimal.

Remark 2.1. We note that the equations (2.1) and (2.2) can be read component-wise: Let

$$M_o(x) := \{d^* \in D(x): r(x,d^*) = \max_{d \in D(x)} (x,d)\}$$

and

$$M_n(x) := \{d^* \in D(x): r(x,d^*) + \sum_{y \in E} p(x,d^*,y)V^*_{n-1}(y)$$

$$= \max_{d \in D(x)} (r(x,d) + \sum_{y \in E} p(x,d,y)V^*_{n-1}(y))\}$$

for every $x \in E$, $n \in \mathbb{N}$. Then $M_n = \prod_{x \in E} M_n(x)$.

All maxima exist by the finiteness assumption with respect to $D(x)$, $x \in E$.

This remark also shows that all sets M_n are non-empty. Hence we obtain

Corollary 2.3. For every $N \in \mathbb{N}_0$ there is an N-step optimal policy in $\mathcal{R}_M$.

Note: We make reference to some further results without discussing them in detail. Analogous results were obtained by HINDERER [49], KUSHNER [75], and SCHÄL [100], [101] under more general assumptions. The asymptotic behaviour of the N-step optimal rewards V_N^* as $N \to \infty$ was studied by BROWN [10], PITTEL [88], HORDIJK [56], HORDIJK, SCHWEITZER and TIJMS [57], and FEDERGRUEN, SCHWEITZER and TIJMS [31]. HINDERER [48] and HINDERER and HÜBNER [53] developed bounds and approximate solutions for N-step problems. An interesting problem is, whether an index $n_0 \in \mathbb{N}_0$ and an $f \in \mathbb{D}$ do exist with $M_n = \{f\}$ for all $n \geq n_0$. Statements concerning this problem are called "turnpike"-theorems. Some of them were given by SHAPIRO [104], DENARDO [14], HINDERER [50], and HINDERER and HÜBNER [52], [53].

THE DISCOUNTED TOTAL REWARD

In the theoretical treatment of the discounted total reward a high level of generality could be achieved. Originally, HOWARD [60] considered the discounted total reward for finite state and decision sets and only for stationary policies. Later on, more general classes of policies (BLACKWELL [8], [9]) and state spaces (BLACKWELL [8], DERMAN [20], MAITRA [81], FURUKAWA [37]) were studied. The finiteness condition with respect to the decision sets was substituted by compactness assumptions, whereas p and r were required to satisfy a continuity condition (GUBENKO and STATLAND [39]). There were investigated also such models whose pay-off structure does not necessarily satisfy a boundedness condition. Since in this case an optimal policy need not exist (DERMAN [20], HORDIJK and TIJMS [59]), generalizations of previous results were given under additional assumptions (HARRISON [43], KÜENLE [74], WESSELS [119]). Finally, decision models of more general classes were considered, e.g., non-stationary models (HINDERER [49]), models with discount factor depending on both the states and decisions (SCHÄL [100], [101]), such admitting inflation (HÜBNER [63], ROTHBLUM [99]), and decision models with a substochastic transition law including the discounted case as a specification (VAN NUNEN [114], PORTEUS [89], ROTHBLUM [99]). Some authors quoted here considered more general models than we do. It seems, however, that our restrictions are hardly restrictive concerning the insight into fundamental results on the discounted case, so they are admitted for the sake of uniformity in this paper.

Now we continue to formulate the results used in chapter 3.

Theorem 2.4. [95]: Let $\beta \in [0,1]$ be a discount factor. The optimal β-discounted total reward V_β^* is the unique solution of the functional equation

$$V_\beta^* = \max_{f \in \mathbb{D}} \{r(f) + \beta P(f) V_\beta^*\}. \qquad (2.3)$$

If f^* is a decision function satisfying

$$f^* \in M_\beta := \{f \in \mathbb{D}: r(f) + P(f)V_\beta^* = V_\beta^*\},$$

then $(f^*)^\infty$ is a β-optimal policy. Furthermore, it holds that $V_\beta^* = V_\beta^{f^*}$.

Remark 2.2. As in theorem 2.1 equation (2.3) can be read component-wise. With

$$M_\beta(x) := \{ d^* \in D(x): r(x,d^*) + \beta \sum_{y \in E} p(x,d^*,y) V_\beta^*(y) =$$

$$= \max_{d \in D(x)} (r(x,d) + \beta \sum_{y \in E} p(x,y,d) V_\beta^*(y))\}$$

for every $x \in E$, we have $M_\beta = \prod_{x \in E} M_\beta(x)$.

As this remark shows it holds that $M_\beta \neq \emptyset$. Hence, we state:

Corollary 2.5. There is a stationary β-optimal policy for every $\beta \in [0,1)$.

Indeed, the theorem 2.4 does not allow us to find a β-optimal policy immediately if the β-optimal total reward V_β^* is not known. In the following we describe a procedure for the step-by-step policy improvement. Due to corollary 2.5 we can confine ourselves to the class of stationary policies.

For every stationary policy f^∞, $f \in \mathbb{D}$, the β-discounted total reward can be represented by

$$V_\beta^f = \sum_{\nu=0}^{\infty} \beta^\nu P(f)^\nu r(f). \qquad (2.4)$$

The series (2.4) converges with respect to the norm topology of $B(E)$ since $\|\beta P(f)\| = \beta < 1$. Hence, the inverse of $I-\beta P(f)$ does exist. By a simple calculation we have, from (2.4),

$$V_\beta^f = (I - \beta P(f))^{-1} r(f). \tag{2.5}$$

For arbitrary $f,g \in \mathbb{D}$ we define real functions $\psi_\beta(f)$ and $\psi_\beta(g,f)$ on K and E, respectively, as follows:

Let

$$\psi_\beta(f)(x,d) := r(x,d) + \sum_{y \in E} (\beta p(x,d,y) - \delta_{xy}) V_\beta^f(y),$$

$(x,d) \in K$, and

$$\psi_\beta(g,f) := \psi_\beta(f)(x,g(x)) \qquad (x \in E).$$

Otherwise stated,

$$\psi_\beta(g,f) = r(g) + (\beta P(g) - I) V_\beta^f.$$

Further, let

$$\mathbb{G}_\beta(f) := \{g \in \mathbb{D}: \psi_\beta(g,f) > 0\}.$$

Theorem 2.6. [95]: For $g,f \in \mathbb{D}$ and $\beta \in [0,1)$, it holds that

(i) $\psi_\beta(g,f) \geqq 0 \Rightarrow V_\beta^g \geqq V_\beta^f$, and

(ii) $g \in \mathbb{G}_\beta(f) \Rightarrow V_\beta^g > V_\beta^f$.

In view of theorem 2.6, the policy improvement procedure can easily be obtained. Let a decision function f be given. Firstly, due to (2.5) the β-discounted total reward V_β^f has to be computed. If the set $\mathbb{G}_\beta(f)$ is non-empty, one chooses a decision function $g \in \mathbb{G}_\beta(f)$ arbitrarily. Due to theorem 2.6 g is "better" than f. The procedure continues with g instead of f. On the other hand, "$\mathbb{G}_\beta(f) = \emptyset$" means that there is no "better" policy than f^∞. Indeed, we obtain:

Theorem 2.7 [95]: If $f \in \mathbb{D}$ is a decision function satisfying $\mathbb{G}_\beta(f) = \emptyset$, then f^∞ is a stationary β-optimal policy.

Notes:

1. If E is finite, so is $\mathbb{D}$. Hence the procedure stops after a finite number of steps having found a β-optimal decision function. In the infinite case the procedure can generate an infinite sequence of policies, whose discounted total rewards converge component-wise to the optimal discounted reward.

2. Usually, the function $g \in \mathbb{G}_\beta(f)$ is chosen as follows: For every state $x \in E$ one selects a decision $d^*(x) \in D(x)$ which maximizes $\psi_\beta(f)(x,d)$. If the maximum vanishes, one defines $d^*(x) = f(x)$. Then g is determined by $g(x) := d^*(x)$, $x \in E$.

2.3. AVERAGE REWARD OPTIMALITY - (P,r)-EQUATIONS

The aim of the following two sections consists in the formulation of a theorem which justifies that the HOWARD procedure [60] for the successive improvement of stationary policies for MDP with several ergodic classes can also be applied in the case of a countably infinite state space. In particular, the aim of the present section consists in investigating the assumptions which have to be required.

It is known from the literature that if state space is finite, there is no distinction between average reward optimality and (-1)-optimality. Since this is not true in the case of infinite state spaces (cf. FLYNN [33], [34]) sensitive discount optimality will be treated in a separate section of this paper. This will turn out to be useful for technical purposes, too.

The results concerning the existence of average optimal policies and the HOWARD algorithm published in the literature are based on conditions which guarantee, roughly speaking, a certain recurrence behaviour of the underlying observation process. If one renounces all the conditions concerning the observation process and the pay-off structure, respectively (in $\mathfrak{R}$ and $\mathfrak{R}_s$, respectively), there need not exist any average optimal policies. This was shown, e.g., by DERMAN [18], FISHER and ROSS [32], MAITRA [80] and ROSS [98].

Analogous statements concerning the applicability are very likely to be made, though the authors do not know whether any related publications already exist. Therefore some "effective" assumptions have to be made in establishing the HOWARD-algorithm for MDP with arbitrary many recurrence classes. The following assumption referring to the decision function $f \in \mathbb{D}$ under consideration plays a central role in the theorem formulated in the next section:

V(f): (i) The transition operator P(f) is 1-recurrent and

(ii) there are functions $g, v \in B(E)$ satisfying the equations

$$g = P(f)g$$

and

$$g + v = r(f) + P(f)v.$$

Condition (i) means that the Markov chain generated by P(f) will attain a positive recurrent state after a (random) time with finite expectation whatever the starting point (cf. lemma 1.8). In the following we discuss the condition (ii).

Firstly, we answer the question how the expected average reward can be represented if one applies a stationary policy f^{∞}, $f \in \mathbb{D}$. An immediate consequence of lemma 1.4 is

Lemma 2.8. Let $f \in \mathbb{D}$ and assume P(f) to be non-dissipative. Then

$$g^f = P(f)^* r(f). \qquad (2.6)$$

Notes:

1. This statement is due to HORDIJK [54]. Since, in the case of a finite space, P(f) is always non-dissipative (2.6) can be applied generally. This fact had been already used by HOWARD [60].

2. The representation (2.6) need not hold if P(f) is dissipative. We demonstrate this by a simple example:
Let $E = \mathbb{N}_0$, $f \in \mathbb{D}$, $r(f) = 1$, and

$$P(f) = \begin{pmatrix} 0 & 1 & 0 & 0 & \dots \\ 0 & 0 & 1 & 0 & \dots \\ 0 & 0 & 0 & 1 & \dots \\ \vdots & \vdots & \vdots & \vdots & \end{pmatrix}$$

("deterministic one-sided walk"). Then, we have $P(f)^n \mathbf{1} = \mathbf{1}$ for $n \in \mathbb{N}_0$ and, hence, $g^f = \mathbf{1}$ on the other hand, $P(f)^* = 0$, thus $P(f)^* r(f) = 0$.

3. As a special consequence of (2.6), g^f is constant on each class of recurrent states (cf. theorem 1.3). Unfortunately, the calculation of g^f is disadvantageous by means of (2.6) since the knowledge of $P(f)^*$ is required. Under certain circumstances g^f can be computed by solving a system of linear equations instead of (2.6). We consider the system

$$g = P(f)g \qquad (2.7)$$

and

$$g + v = r(f) = P(f)v \qquad (2.8)$$

in real functions g and v on E.

Definition 2.17. The equations (2.7) and (2.8) are called (P,r)-equations (with respect to f). If real functions g and v solve the system (2.7)/(2.8), both the pair (g,v) and g and v, respectively, are called a solution of the (P,r)-equations; it is called bounded if g and v are bounded.

It is possible that the (P,r)-equations do not have any solution (cf. example 2.1). Therefore, the existence of a bounded solution must be assumed in several statements.

Lemma 2.9. Let $f \in \mathbb{D}$, and let (g,v) be a bounded solution of the (P,r)-equations (with respect to f). Then $g^f = g$.

Proof: Applying $P(f)^n$, $n \in \mathbb{N}_0$, to both sides and after re-ordering one obtains that

$$g = P(f)^n r(f) + P(f)^{n+1} v - P(f)^n v.$$

A summation of these equations with respect to n (n = 1,...,N-1) yields

$$Ng = \sum_{n=0}^{N-1} P(f)^n r(f) + (P(f)^N - I)v.$$

Based on the inequality

$$\|(P(f)^N - I)v\| \leq 2\|v\|,$$

which is valid for every $N \in \mathbb{N}_0$, it follows that

$$g = \lim_{N\to\infty} \{\frac{1}{N} \sum_{n=0}^{N-1} P(f)^n r(f) + \frac{1}{N}(P(f)^N - I)v\}$$

$$= \lim_{N\to\infty} \frac{1}{N} \sum_{n=0}^{N-1} P(f)^n r(f) = g^f. \blacksquare$$

Using lemma 2.9 one can see that, if the (P,r)-equations have a bounded solution, the function g is uniquely defined. In contrast, v belongs to a manifold of solutions.

Lemma 2.10. Let $f \in \mathbb{D}$, let P(f) be non-dissipative, and let the (P,r)-equations (with respect to f) have a bounded solution. Then the solution (g,v) of the (P,r)-equations has the general form $g = g^f$ and $v \in v_o + \ker(I - P(f)^*)$[1], where $v_o \in B(E)$ is an arbitrary particular solution.

Proof: In view of the assumption, there is a special solution (g, v_o) in B(E) which satisfies $g = g^f$ (lemma 2.9). Now it is **sufficient to show** that $v \in v_o + \ker(I - P(f))$ since, due to theorem 1.7,

$$\ker(I - P(f)^*) = \ker(I - P(f)).$$

By definition, for every $v_1 \in \ker(I - P(f))$ it holds that $v_1 = P(f)v_1$. Therefore, it follows from (2.8) that

$$g + v_o + v_1 = r(f) + P(f)v_o + v_1 = r(f) + P(f)(v_o+v_1).$$

Thus (v_o+v_1) is also a solution of the (P,r)-equations. Now we consider the opposite case. Let v_2 be any solution of (2.8). By subtraction of

$$g^f + v_o = r(f) + P(f)v_o$$

from

$$g^f + v_2 = r(f) + P(f)v_2,$$

we obtain

$$(v_2-v_o) = P(f)(v_2-v_o),$$

i.e., $v_2 \in v_o + \ker(I - P(f))$. $\blacksquare$

Corollary 2.11. If, for an arbitrary $f \in \mathbb{D}$, the matrix P(f) is non-dissipative and the (P,r)-equations (with respect to f) have a bounded solution, then there is a uniquely determined solution (g^f, v_o) satis-

---

[1]) $v_o + \ker(I - P(f)^*) = \{v_o + v: v \in \ker(I - P(f)^*)\}$

fying

$$P(f)^* v_0 = 0. \qquad (2.9)$$

Proof: Let (v, g^f) be any particular solution. Then $P(f)^* v_1 \in \ker(I - P(f)^*)$ since $P(f)^*$ is idempotent due to lemma 1.4. Hence, in view of lemma 2.10,

$$v_0 := v_1 - P(f)^* v_1$$

is also a solution of (2.8). Obviously,

$$P(f)^* v_0 = 0.$$

Now, let (g^f, v_0') be a bounded solution of (2.7) - (2.9). By lemma 2.10 and (2.9), we obtain that

$$v_0 - v_0' = P(f)^*(v_0 - v_0') = 0. \blacksquare$$

In view of the preceding lemmas it is reasonable to define:

Definition 2.18. Let $f \in D$, and let $P(f)$ be non-dissipative. Then the equations

$$g = P(f)g \qquad (2.7)$$

$$g + v = r(f) + P(f)v \qquad (2.8)$$

$$P(f)^* v = 0$$

in bounded real functions g and v on E are called $(P,r)^*$-equations (with respect to f). If these equations possess a solution, then it is denoted by (g^f, v^f).

Note: First of all, HOWARD [60] considered the (P,r)-equations and proved the assertions of the lemmas 2.9 and 2.10, where the state space was assumed to be finite. Moreover, it was shown that the (P,r)-equations are solvable in this case; of course, each solution must be bounded. Hence, in the finite state case, the $(P,r)^*$-equations are always solvable. (The latter statement was given, e.g., by SLADKY [106]).

Remark 2.3. Not only g^f but also $\mathfrak{v}^f$ allows a substantial interpretation as it was shown, e.g., by HOWARD [60]. Under the condition that E is finite and $P(f)$ is aperiodic he showed that, for a stationary policy f^∞, $f \in \mathbb{D}$, the N-stage rewards v_N^f can be asymptotically represented in the form

$$v_N^f = N g^f + v^f + o(1),$$

where v^f is a so-called "weight vector". Now, it is easy to show that $v^f = \mathfrak{v}^f$ (cf. DENARDO [14], who called $\mathfrak{v}^f$ the "bias" of the policy f^∞). Another interpretation of $\mathfrak{v}^f$ will be given in section 2.5.

Remark 2.4. The (P,r)-equations can be essentially simplified if, for a non-dissipative transition matrix P(f), the underlying Markov chain has only one class of recurrent states. Since in this case 1 is a simple eigenvalue of P(f) in B(E) (cf. theorem 1.7) g^f can be written as $g^f = g^f \mathbf{1}$ with a suitable constant $g^f \in \mathbb{R}$. (2.7) and (2.8) change into

$$g^f \mathbf{1} + v = r(f) + P(f)v. \qquad (2.8')$$

Note: The boundedness of g seems to be a necessary condition for proving the identity $g = g^f$ for any solution (g,v) of the (P,r)-equation (with respect to f, $f \in \mathbb{D}$) (cf. [123], [75]), this does not hold for the boundedness of v (cf. theorem 2.15). In the present paper, we confine our interest to bounded solutions since they are of particular significance with respect to the HOWARD algorithm.

In the following we give some sufficient conditions guaranteeing the (P,r)-equations to have a bounded solution. For this purpose we use the notion of a "strong Markov chain" (cf. definition 1.8).

Theorem 2.12. (ASO and KIMURA [1]) Let $f \in \mathbb{D}$, and let P(f) be a strong Markov chain. Then the (P,r)-equations (with respect to f) have a bounded solution.

Corollary 2.13. Let $f \in \mathbb{D}$. If P(f) is quasi-compact, then the (P,r)-equations (with respect to f) have a bounded solution.

Proof: Due to theorem 1.20, P(f) is a strong Markov chain. Hence, theorem 2.12 can be applied. ∎

In view of a further specification of the assumptions, we obtain the following known result.

Corollary 2.14. (DERMAN and VEINOTT [23]) Let $f \in \mathbb{D}$, and let there exist a constant $a \in \mathbb{R}$ and a state $x_0 \in E$ such that for every $x \in E$

$$E_x^f \tau_{x_0} \leq a.$$

Then there exists a bounded solution $(g\,\mathbf{1}, v)$ $(g \in \mathbb{R})$ of the equation

$$g\mathbf{1} + v = P(f)v + r(f) \qquad (2.8').$$

Subsequently, an additional result due to WIJNGAARD [123] concerning the existence of (not necessarily bounded) solutions of the (P,r)-equations will be mentioned. It can be re-formulated to an existence theorem concerning the bounded solutions of the (P,r)-equations by applying slight modifications.

Let $A \subset E$ be arbitrary, let $B := E \setminus A$ and let P be a transition matrix. Further, let ${}_AP$ denote the substochastic $E \times E$-matrix defined by

$${}_AP(x,y) := P(x,y)1_A(y), \quad x,y \in E.$$

WIJNGAARD [123] showed that a substochastic $E \times E$-matrix P_A is defined by

$$(x,y) \to P_A(x,y) = \sum_{n=0}^{\infty} ((_BP)^n \, _AP1_{\{y\}})(x),$$

$(x,y \in E)$. The matrix P_A is stochastic, if and only if, for every $x \in E$,

$$\lim_{n\to\infty}((_BP)^n 1)(x) = 0.$$

In this case P_A is called the embedded Markov chain of P in A. Let us remark that it is not a Markov chain in the sense of definition 1.2.

Definition 2.19. Let $A \subset E$, and let P be a transition matrix on E. P is called $(A, \mathbf{1})$-recurrent if (with $B := E \setminus A$)

(i) the series $\sum_{m=0}^{\infty} ((_BP)^m \mathbf{1})(x)$ converges for every $x \in E$,

(ii) this convergence is uniform with respect to x on A, and

(iii) the function defined by

$$x \to s(x) = \sum_{m=0}^{\infty} ((_BP)^m \mathbf{1})(x)$$

is bounded on A.

P is called strongly $(A, \mathbf{1})$-recurrent if P is $(A, \mathbf{1})$-recurrent and if s is bounded on E.

Theorem 2.15. (WIJNGAARD [123]) Let $f \in \mathbb{D}$, and let $A \subset E$ be a set such that P(f) is $(A, \mathbf{1})$-recurrent and the embedded Markov chain $P(f)_A$ is quasi-compact. Then the (P,r)-equations (with respect to f) have a solution (g_0, v_0) satisfying

$$g_0 = g^f \text{ and } \lim_{m\to\infty} \frac{1}{m} P(f)^m v_0 = 0$$

(in particular, $P(f)^m v_0$ does exist for every $m \in \mathbb{N}$).

Without proof we remark that, by a modification of the proof given in [123], from theorem 2.15 the following result can be obtained.

Theorem 2.16. Let $f \in \mathbb{D}$, and let A ($A \subset E$) be a set such that P(f) is strongly $(A, \mathbf{1})$-recurrent and the embedded Markov chain $P(f)_A$ of P on A is quasi-compact. Then the (P,r)-equations (with respect to f) have a bounded solution.

Note: Hence, corollary 2.14 is both a special case of theorem 2.16 (since P(f) is strongly $(\{x_0\}, \mathbf{1})$-recurrent) and of corollary 2.13, because (2.10) implies the DOEBLIN condition. Thus P(f) is quasi-compact.

DISCUSSION OF THE ASSUMPTION V(f):

After sufficient conditions concerning the solvability of the $(P,r)^*$-equations have been established, the difference between part (i) and part (ii) of the assumption V(f) remains to be clarified and some important cases in which V(f) is fulfilled must be turned out.

Firstly, we consider the question whether, at the same time, 1-recurrence of a transition matrix P(f), $f \in \mathbb{D}$, implies that the $(P,r)^*$-equations are solvable and whether a reversed implication is valid, respectively.

Example 2.1. Let $E = \mathbb{N}_0$, and let $(a_n)_{n\in\mathbb{N}}$ be a strictly increasing sequence of numbers $0 < a_n < 1$, $n \in \mathbb{N}$, with $\lim_{n\to\infty} a_n = 1$. For a certain $f \in \mathbb{D}$, assume that P(f) has the form

$$P(f) = \begin{pmatrix} 1 & 0 & 0 & 0 & \cdots \\ 1-a_1 & a_1 & 0 & 0 & \cdots \\ 1-a_2 & 0 & a_2 & 0 & \cdots \\ 1-a_3 & 0 & 0 & a_3 & \cdots \\ \vdots & \vdots & \vdots & \vdots & \end{pmatrix},$$

and let

$$r(f) = \begin{pmatrix} 1 \\ -1 \\ 1 \\ -1 \\ \vdots \end{pmatrix}.$$

For $n \in \mathbb{N}$

$$P(f)^n = \begin{pmatrix} 1 & 0 & 0 & 0 & \cdots \\ 1-a_1^n & a_1^n & 0 & 0 & \cdots \\ 1-a_2^n & 0 & a_2^n & 0 & \cdots \\ 1-a_3^n & 0 & 0 & a_3^n & \cdots \\ \vdots & \vdots & \vdots & \vdots & \end{pmatrix}.$$

Thus (component-wise):

$$P(f)^* = \lim_{n\to\infty} P(f)^n = \begin{pmatrix} 1 & 0 & 0 & \cdots \\ 1 & 0 & 0 & \cdots \\ 1 & 0 & 0 & \cdots \\ \vdots & \vdots & \vdots & \end{pmatrix},$$

and $P(f^*)\,\mathbf{1} = \mathbf{1}$, i.e., P(f) is non-dissipative. Furthermore, $T := \mathbb{N}$ is exactly the set of transient states of P(f). It holds that

$$(P(f)_{TT})^n = \begin{pmatrix} a_1^n & 0 & 0 & \cdots \\ 0 & a_2^n & 0 & \cdots \\ 0 & 0 & a_3^n & \cdots \\ \vdots & \vdots & \vdots & \end{pmatrix}$$

for $n \in \mathbb{N}$. Therefore, if $x \in T$, then

$$(\Pi\, \mathbf{1}_T)(x) := \Big(\sum_{n=0}^{\infty} (P(f)_{TT})^n\, \mathbf{1}_T\Big)(x) = \frac{1}{1-a_x} < \infty\,.$$

Thus it was shown that P(f) is 1-recurrent. If we consider the (P,r)-equations, then each solution g of (2.7) must be of the form $g = g_0\,\mathbf{1}$ with a certain constant $g \in \mathbb{R}$. According to remark 2.4 only the system

$$g_0\,\mathbf{1} + v = r(f) + P(f)v \qquad (2.8')$$

has to be solved. It can be rewritten to

$$(I - P(f))v = r(f) - g_0\,\mathbf{1} =: h$$

or

$$\begin{pmatrix} 0 & 0 & 0 & 0 & \cdots \\ a_1-1 & 1-a_1 & 0 & 0 & \cdots \\ a_2-1 & 0 & 1-a_2 & 0 & \cdots \\ a_3-1 & 0 & 0 & 1-a_3 & \cdots \\ \vdots & \vdots & \vdots & \vdots & \end{pmatrix} \begin{pmatrix} v(0) \\ v(1) \\ v(2) \\ v(3) \\ \vdots \end{pmatrix} = \begin{pmatrix} h(0) \\ h(1) \\ h(2) \\ h(3) \\ \vdots \end{pmatrix}.$$

From this we obtain

$$v(2n+1) = \frac{-1-g_0}{1-a_n} + v(0), \quad n \in \mathbb{N}_0,$$

and

$$v(2n) = \frac{1-g_0}{1-a_n} + v(0), \qquad n \in \mathbb{N}_0.$$

Consequently, it is obvious that, whatever the choice of g_0 and v(0),

$$|v(n_k)| \to \infty \quad (k \to \infty)$$

at least for a certain subsequence $(v(n_k))_{k\in N}$. Hence, no bounded solution of (2.8') can exist.

Example 2.1 shows that from the 1-recurrence we cannot conclude that the $(P,r)^*$-equations have a solution. (Further, we note that the matrix P(f) of the example is (A,1)-recurrent with $A := \{0\}$. Moreover, since obviously $\hat{P}(f)_A = 1$, the assumption of theorem 2.15 are fulfilled. However, because there is no bounded solution of the (P,r)-equations the assumptions of theorem 2.16 cannot be fulfilled. Thus the assumptions properly strengthen those of theorem 2.15). The question whether the transition matrix under consideration is 1-recurrent must be negatively answered, too.

The following

Example 2.2. shows a transient Markov chain whose (P,r)-equations possess a bounded solution:

Let $f \in \mathbb{D}$, $E = \mathbb{N}_0$ and

$$P(f) = \begin{pmatrix} 0 & 1 & 0 & 0 & \cdots \\ 0 & 0 & 1 & 0 & \cdots \\ 0 & 0 & 0 & 1 & \cdots \\ \vdots & \vdots & \vdots & \vdots & \end{pmatrix}, \quad r(f) = \begin{pmatrix} 1 \\ 1/2 \\ 1/4 \\ \vdots \end{pmatrix},$$

i.e., $r(f)(x) = 2^{-x}$ for $x \in \mathbb{N}_0$. Every solution of (2.7) has the form $g = g_0 \mathbf{1}$ with a certain constant $g_0 \in \mathbb{R}$.

Let $h(x) := r(f)(x) - g_0$ for $x \in \mathbb{N}_0$. The system of equations (2.8) is given by

$$(I - P(f))v = r(f) - g_0 \mathbf{1} = h$$

or

$$\begin{pmatrix} 1 & -1 & 0 & 0 & \cdots \\ 0 & 1 & -1 & 0 & \cdots \\ 0 & 0 & 1 & -1 & \cdots \\ \vdots & \vdots & \vdots & \vdots & \end{pmatrix} \begin{pmatrix} v(0) \\ v(1) \\ v(2) \\ \vdots \end{pmatrix} = \begin{pmatrix} h(0) \\ h(1) \\ h(2) \\ \vdots \end{pmatrix}.$$

Thus we obtain

$$v(n) = v(0) - \sum_{k=0}^{n-1} h(x), \quad n \in \mathbb{N},$$

where v(0) can be arbitrarily chosen. Hence, for example, $g := 0$ and v defined by v(0), $v(n) := 2(1-2^{-n})$, $n \in \mathbb{N}$, provide the existence of a bounded solution of the (P,r)-equations.

After having shown that the two parts (i) and (ii) of the assumption V(f) do not imply each other, we now consider the relation between V(f) and the assumptions of the theorem 2.12 and the corollaries 2.13/14. The assumptions of these statements imply that the transition matrix under consideration is a strong Markov chain. Since this yields the 1-recurrence of P, it is clear that the assumptions of the statements 2.12-14 imply the validity of V(f). Now, we show that this statement cannot be reversed.

Example 2.3. Let $E = \mathbb{N}_0$, $f \in \mathbb{D}$, and let P(f) be defined as in example 2.1. Furthermore, let $r(f)(x) := a_x$ for $x \in E$. In the same way as in example 2.1, the (P,r)-equation can be transformed into the system of equations

$$v(n) = \frac{a_n - g_0}{1 - a_n} + v(0), \quad n \in \mathbb{N},$$

in real constants g_0 and $v(n)$, $n \in \mathbb{N}$. Obviously, $g = g_0 \mathbf{1}$ with $g_0 = 1$ and v defined by $v(0) = 0$ and $v(n) = -1$ ($n \in \mathbb{N}$) provide a bounded solution of the (P,r)-equations. Since it was already shown that

$$\Pi \mathbf{1}_T = (1 - a_x)^{-1} \quad (x \in T := E \setminus \{0\})$$

and consequently $\lim_{x\to\infty} \Pi \mathbf{1}_T = \infty$, the submatrix $P(f)_{TT}$ of P(f) cannot be strongly absorbable, i.e., P(f) is not a strong Markov chain.

In view of this example it can be seen that the assumption V(f) is, in fact, weaker than the assumption that P(f) is a strong Markov chain. Simultaneously, it turns out that the solvability of the (P,r)*-equations does not only depend on the recurrence properties of the transition operator but also on the structure of the pay-off function.

2.4. A HOWARD TYPE ALGORITHM

Now we attain a main result of this chapter. For this purpose we first introduce some definitions.

Let $g, f \in \mathbb{D}$ be decision functions whose (P,r)-equations are solvable in B(E). Define real functions $\phi^{-1}(f)$ and $\phi^{0}(f)$, $\phi^{-1}(g,f)$ and $\phi^{0}(g,f)$, respectively, on K and E, respectively, as follows:
Let

$$\phi^{-1}(f)(x,d) := \sum_{y \in E} (p(x,d,y) - \delta_{xy}) g^f(y) \quad (x,d) \in K$$

and

$$\phi^{0}(f)(x,d) := r(x,d) + \sum_{y \in E} (p(x,y,d) - \delta_{xy}) v^f(y) - g^f(x).$$

Furthermore, let

$$\phi^{-1}(g,f)(x) := \phi^{-1}(f)(x,g(x)) \quad (x \in E)$$

and

$$\phi^{0}(g,f)(x) := \phi^{0}(f)(x,g(x)).$$

Hence, we obtain

$$\psi^{o}(g,f) = r(g) + (P(g) - I)v^{f} - g^{f}.$$

These functions can **rewritten as vector-valued functions** $\Psi^{o}(f)$ **and** $\Psi^{o}(g,f)$ on K and E, respectively, in the following way:
For $(x,d) \in K$ let

$$\psi^{o}(f)(x,d) := (\psi^{-1}(f)(x,d),\ \psi^{o}(f)(x,d)),$$

and let

$$\psi^{o}(g,f)(x) := (\psi^{-1}(g,f)(x),\ \psi^{o}(g,f)(x)).$$

This implies

$$\psi^{o}(f) = (\psi^{-1}(f),\ \psi^{o}(f)) \text{ and } \psi^{o}(g,f) = (\psi^{-1}(g,f),\ \psi^{o}(g,f))$$ [1]

Definition 2.20. A state $x \in E$ is called recurrent (and transient, respectively) under g ($g \in \mathbb{D}$) if x is a recurrent (and transient, respectively) state with respect to $P(g)$. For any $x, x' \in E$, x' is called accessible from x under g if x' is an accessible state from x with respect to $P(g)$. In this case we briefly write "$x \overset{g}{\curvearrowright} x'$".

Now we consider the following assertions for an arbitrary state $x \in E$:

$A(x)$: $\Longleftrightarrow$ "There is a recurrent state x' under g with $x \overset{g}{\curvearrowright} x'$ and $\psi^{o}(g,f)(x') > 0$."

$B(x)$: $\Longleftrightarrow$ "There is a state x'' satisfying $x \overset{g}{\curvearrowright} x''$ and $\psi^{-1}(g,f)(x'') > 0$."

$C(x)$: $\Longleftrightarrow$ "For every state x' being recurrent under g with $x \overset{g}{\curvearrowright} x'$, it holds that $\psi^{o}(g,f)(x') = 0$, and for each $x'' \in E$ with $x \overset{g}{\curvearrowright} x''$ the equation $\psi^{-1}(g,f)(x'') = 0$ is valid."

Clearly, $C(x) \Longleftrightarrow \neg(A(x) \vee B(x))$. Furthermore, let

$C^{*}(x)$: $\Longleftrightarrow$ "$C(x)$ is valid, and there is a state x' satisfying $x \overset{g}{\curvearrowright} x'$ and $\psi^{o}(g,f)(x') > 0$."

Using these abbreviations we can formulate the following main theorem.

Theorem 2.17. Let g and f be decision functions whose $(P,r)^{*}$-equations are solvable. Let $P(g)$ be 1-recurrent, and let $\Psi^{o}(g,f) \succ 0$. [2]

(i) Then $g^{g} \geq g^{f}$.

For each $x \in E$

(ii) $A(x) \vee B(x) \Longleftrightarrow g^{g}(x) > g^{f}(x)$, and

(iii) $C(x) \Longleftrightarrow g^{g}(x) = g^{f}(x)$.

---

[1]) Since the different mathematical objects $\psi^{o}(f)$ and $\psi^{o}(g,f)$ are in close touch with another, we use the unified notation Ψ^{o}.

[2]) "$\succ$" denotes the lexicographical order (cf. A1)

If g satisfies the additional assumption

$$\psi^o(g,f)(x) = 0 \Rightarrow g(x) = f(x) \qquad (2.10)$$

for $x \in E$ we additionally obtain the equivalence relations:

(iv) $C(x) \Leftrightarrow g^g(x) = g^f(x) \wedge v^g(x) \geq v^f(x)$, and

(v) $\dot{C}^*(x) \Leftrightarrow g^g(x) = g^f(x)$, $v^g(x) > v^f(x)$, and x is transient under g.

Proof: (For brevity, we write ψ^{-1} instead of $\psi^{-1}(g,f)$, and ψ^o instead of $\psi^o(g,f)$.)

It holds that

$$\begin{aligned}\Delta g := g^g - g^f &= P(g)g^g - g^f \quad \text{(by (2.7))}\\ &= (P(g) - I)g^f + P(g)(g^g - g^f)\\ &= (P(g) - I)g^f + P(g)\Delta g\ ,\end{aligned}$$

that is

$$\Delta g = \psi^{-1} + P(g)\Delta g. \qquad (1^*)$$

Applying the bounded operator $P(g)^*$ to both sides of this equation (1*) we see that

$$\begin{aligned}P(g)^*\Delta g &= P(g)^*\psi^{-1} + P(g)^*(P(g)\Delta g)\\ &= P(g)^*\psi^{-1} + (P(g)^*P(g))\Delta g\\ &= P(g)^*\psi^{-1} + P(g)^*\Delta g\ ,\end{aligned}$$

since

$$P(g)^*P(g) = P(g)^*$$

(cf. lemma 1.3).

Hence,

$$P(g)^*\psi^{-1} = 0. \qquad (2^*)$$

The 1-recurrence assumption with respect to P(g) implies that P(g) is non-dissipative. Therefore, $P(g)^*(x,y) > 0$, if and only if y is recurrent under g, and $x \overset{g}{\rightsquigarrow} y$ (cf. theorem 1.3). Let R denote the set of states being recurrent under g, and $T := E \setminus R$. Then, from (2*), by $\Psi^o(g,f) \succ 0$,

$$\psi_R^{-1} = 0 \text{ and } \psi_R^o \geq 0. \qquad (3^*)$$

Suppose that $R_1, R_2, \ldots, R_n, \ldots$ are the recurrence classes under g, i.e., $R = \bigcup_n R_n$. By (3*) and lemma 1.5, (1*) changes into

$$\Delta g_{R_n} = P(g)\Delta g|_{R_n} = P(g)_{R_nR_n}\,\Delta g_{R_n} \quad (4^*)$$

for all n, and into

$$\Delta g_T = \psi_T^{-1} + P(g)\Delta g|_T = \psi_T^{-1} + \sum_n P(g)_{TR_n}\Delta g_{R_n} + P(g)_{TT}\,\Delta g_T\ . \quad (5^*)$$

Following theorem 1.6 and (4*) there is a constant $c_n \in \mathbb{R}$ for every n which satisfies

$$\Delta g_{R_n} = c_n \mathbf{1}_{R_n}.$$

(5*) can be re-ordered into

$$(I - P(g)_{TT})\Delta g_T = \psi_T^{-1} + \sum_n P(g)_{TR_n} \Delta g_{R_n},$$

whose left-hand part $(I - P(g)_{TT})\Delta g_T$ is bounded. Since $P(g)_{TT}$ is 1-absorbable, both sides can be multiplied by the matrix

$$\Pi := \sum_{k=0}^{\infty} (P(g)_{TT})^k,$$

which leads to

$$\Pi(I - P(g)_{TT})\Delta g_T = \Pi(\psi_T^{-1} + \sum_n P(g)_{TR_n} \Delta g_{R_n}).$$

However, by theorem 1.9

$$\Pi(I - P(g)_{TT})\Delta g_T = \Delta g_T ;$$

hence,

$$\Delta g_T = \Pi(\psi_T^{-1} + \sum_n P(g)_{TR_n} \Delta g_{R_n}). \qquad (6^*)$$

Now, we consider the difference $\Delta v := v^g - v^f$. Due to (2.8),

$$\begin{aligned}\Delta g + \Delta v &= g^g + v^g - g^f - v^f \\ &= r(g) + P(g)v^g - r(f) - P(f)v^f \\ &= r(g) + (P(g) - I)v^f - (r(f) + (P(f) - I)v^f) \\ &\quad + P(g)(v^g - v^f),\end{aligned}$$

and, accepting the equality

$$r(f) + (P(f) - I)v^f = g^f ,$$

which follows from (2.8), that means that

$$\Delta g + \Delta v = \psi^o + P(g)\Delta v. \qquad (7^*)$$

The application of $P(g)^*$ on both sides of (7*) yields

$$P(g)^*\Delta g = P(g)^*\psi^o$$

or, rewritten for the individual recurrence classes R_n,

$$(P(g)_{R_nR_n})^*\psi^o_{R_n} = (P(g)_{R_nR_n})^*\Delta g_{R_n} = c_n \mathbf{1}_{R_n}.$$

Therefore, (3*) implies $\Delta g_{R_n} \geq 0$ for every n. For every $x \in R$

$$(\Delta g)(x) > 0 \Leftrightarrow \exists x' : x \overset{g}{\rightsquigarrow} x', \psi^o(x') > 0, x' \text{ is recurrent under } g, \qquad (8^*)$$

since all elements of each matrix $(P(g)_{R_nR_n})^*$ are positive.

The matrix Π obviously fulfils the condition that $\Pi(x,y) \geq 0$ for $x,y \in T$, and

$$\Pi(x,y) > 0 \iff x \overset{g}{\curvearrowright} y.$$

Now, let $x \in E$ be transient under g. Then it follows from (6*) via (8*) that

$$(\Delta g)(x) > 0 \iff (\exists x' \in E\colon x \overset{g}{\curvearrowright} x',\ \phi^{-1}(x') > 0) \vee (\exists x'' \in R\colon x \overset{g}{\curvearrowright} x'',\ \phi^{0}(x'') > 0). \qquad (9^*)$$

In particular, $\Delta g_T \geqq 0$ since $\phi_T^{-1} \geqq 0$ and since all Δg_{R_n} are non-negative **as we showed above. This proves (i).**

The equivalence relation (8*) is identical with

$$(\Delta g)(x) > 0 \iff A(x)$$

for every recurrent state x under g. Noting that B(x) is false for every $x \in R$, and that (9*) is equal to

$$(\Delta g)(x) > 0 \iff A(x) \vee B(x)$$

for $x \in T$, it is possible to reformulate (8*) and (9*) into the equivalence relation asserted in (ii), which is valid for all $x \in E$.

(iii) is a trivial consequence of (ii) due to the logical argument that

$$C(x) \iff \neg\,(A(x) \vee B(x)).$$

Now, we consider the situation described in (iv) and (v), respectively. For $x \in E$, let

$$T(x) := \{y \in T\colon x \overset{g}{\curvearrowright} y\} \text{ and } R(x) := \{y \in R\colon x \overset{g}{\curvearrowright} y\}.$$

If equation (7*) is analogously decomposed as was done with the equation (1*), we obtain

$$\Delta g_{R_n} + \Delta v_{R_n} = \phi^{0}_{R_n} + P(g)_{R_n R_n} \Delta v_{R_n} \qquad (10^*)$$

for all n, and

$$\Delta g_T + \Delta v_T = \phi^{0}_T + \sum_n P(g)_{TR_n} \Delta v_{R_n} + P(g)_{TT} \Delta v_T .$$

The latter can be rewritten as

$$\Delta v_T = \Pi(\phi^{0}_T - \Delta g_T + \sum_n P(g)_{TR_n} \Delta v_{R_n}). \qquad (11^*)$$

Now, we suppose that $x \in E$ is a recurrent state under g. Then $R(x) = R_1$ can be assumed without loss of generality. If C(x) is valid, then

$$\phi^{-1}_{R_1} = \phi^{0}_{R_1} = 0,$$

and (10*) changes to

$$\Delta v_{R_1} = P(g)_{R_1 R_1} \Delta v_{R_1} .$$

Thus there is a constant $c_1 \in \mathbb{R}$ with $\Delta v_{R_1} = c_1 \mathbf{1}_{R_1}$. Furthermore, due to (2.10) we have $g_{R_1} = f_{R_1}$, implying that

$$(P(g)_{R_1R_1})^* = (P(f)_{R_1R_1})^*$$

by lemma 1.5. Now (2.9) yields

$$\begin{aligned}(P(g)_{R_1R_1})^*\Delta v_{R_1} &= (P(g)_{R_1R_1})^*(v^g_{R_1} - v^f_{R_1}) \\ &= (P(g)_{R_1R_1})^* v^g_{R_1} - (P(f)_{R_1R_1})^* v^f_{R_1} \\ &= (P(g)^* v^g - P(f)^* v^f)|_{R_1} = 0.\end{aligned}$$

Thus

$$\Delta v_{R_1} = c_1 \mathbf{1}_{R_1} = (P(g)_{R_1R_1})^* \ c_1 \mathbf{1}_{R_1} = 0.$$

With respect to (iii) these considerations show that the equivalence relation

$$C(x) \Longleftrightarrow g^g(x) = g^f(x) \wedge v^g(x) = v^f(x) \qquad (12^*)$$

holds for every state x being recurrent under g.

Let us now investigate the opposite case of x being transient under g. If C(x) holds (i.e., $g^g(x) = g^f(x)$), then we immediately obtain $\psi^{-1}_{T(x)\cup R(x)} = 0$ and $\psi^o_{R(x)} = 0$.

By $\psi^o(g,f) \succcurlyeq 0$, it follows $\psi^o_{T(x)} \geqq 0$. With respect to the properties of the matrix Π mentioned above, (11*) changes to

$$(\Delta v_T)(x) = (\Pi_{TT(x)}(\psi^o_{T(x)} - \Delta g_{T(x)} + \sum_n P(g)_{T(x)R_n} \Delta v_{R_n}))(x).$$

In the following we show that, moreover, C(x) implies the strengthened equation

$$(\Delta v_T)(x) = (\Pi_{TT(x)} \ \psi^o_{T(x)})(x). \qquad (13^*)$$

Firstly, $\Delta g_{T(x)} = 0$ is valid. Otherwise, one could find a state $x_o \in T(x)$ satisfying $(\Delta g_{T(x)})(x_o) > 0$. In this case, $A(x_o)$ or $B(x_o)$ should be valid. Since $x \overset{g}{\rightarrow} x_o$, this would imply $A(x) \vee B(x)$ – a contradiction.

Furthermore,

$$\sum_n P(g)_{T(x)R_n} \Delta v_{R_n} = 0.$$

This can be proved in the following way:
Let n be fixed. Then either $P(g)(y,R_n) = 0$ for all $y \in T(x)$, implying $P(g)_{T(x)R_n} \Delta v_{R_n} = 0$, or there is an $x_o \in T(x)$ satisfying $P(g)(x_o,R_n) > 0$. In the latter case we have $x \overset{g}{\rightarrow} y$ for all $y \in R_n$, that is, C(y) holds for all $y \in R_n$. Having proved this assertion we obtain $\Delta v_{R_n} = 0$; hence $P(g)_{T(x)R_n} \Delta v_{R_n} = 0$. Thus (13*) results. This implies that $(\Delta v_T)(x) \geqq 0$, and, consequently, (iv) is proved.

Due to (12*) and (13*), it is obvious that, under C(x), $(\Delta v_T)(x) > 0$, if and only if $\psi^{o}_{T(x)} > 0$. However, the latter is only possible if x is transient under g. Thus (v) results, and the proof is complete. ∎

The importance of the theorem just proved consists in its inherent possibility to apply a policy improvement procedure for stationary policies. In addition, several well-known results of other authors can be obtained from theorem 2.17 by specification, as we shall show later on. Above all, we shall prove two simple corollaries.

Corollary 2.18. Let f,g be decision functions whose $(P,r)^*$-equations are solvable. Furthermore, let P(g) be 1-recurrent, and $\Psi^{o}_{T(x)} \succcurlyeq 0$. Then, for each $x \in E$,

(i) $\psi^{-1}(g,f)(x) > 0 \Rightarrow g^g(x) > g^f(x)$.

In addition, if x is recurrent under g, we have

(ii) $\psi^{o}(g,f)(x) > 0 \Rightarrow g^g(x) > g^f(x)$.

Proof: If $\psi^{-1}(g,f)(x) > 0$, then B(x) is valid. $\psi^{o}(g,f)(x) > 0$ implies A(x). Hence (i) and (ii) are conclusions of the theorem 2.17 (ii).

Remarks:

2.5. If E is finite, for every $f \in \mathbb{D}$, P(f) is 1-recurrent, and the $(P,r)^*$-equations have a solution. That is, $\Psi^{o}(g,f) \succcurlyeq 0$ generally implies $g^g \geqq g^f$. This assertion is due to HOWARD [60]. Under the same condition, part (ii) of the above corollary was proved by VEINOTT [117].

2.6. Consider the following assumptions:

(A1): For every $f \in \mathbb{D}$, P(f) is irreducible and positive recurrent.

(A2): For every $f \in \mathbb{D}$, the $(P,r)^*$-equations have a solution with the property that $g^f = g\,1$ for a certain constant $g^f \in \mathbb{R}$.

On condition that (A1) and (A2) are fulfilled, theorem 2.17 and corollary 2.18, respectively, can be applied since, by (A1), P(f) must be 1-recurrent for every $f \in D$. Then part (ii) of corollary 2.18 has the form:

$$\psi^{o}(g,f) > 0 \Rightarrow g^g > g^f.$$

This special case was shown by DERMAN [18].

2.7. KUSHNER [75] introduced the following conditions:

(A3): For every $f \in \mathbb{D}$ and $x \in E$, the series

$$\sum_{n=0}^{\infty} (P(f)^n - P(f)^*)r(f)(x) =: s^f(x)$$

is convergent and uniformly bounded in $f \in \mathbb{D}$.

(A4): On condition that (A3) is fulfilled, $s^f(x)$ is uniformly bounded in $x \in E$.

[75] shows that the three conditions (A1), (A3), and (A4) as a whole imply (A2). In this case s^f is equal to v^f. Now KUSHNER proved the

following:
If, under the conditions (A1), (A3), and (A4), a certain $g^* \in \mathbb{D}$ fulfils the equation

$$\Psi^o(g^*,f) = \max_{g \in \mathbb{D}} \Psi^o(g,f),$$

then $g^{g^*} \geqq g^f$. This is a special case of part (i) of theorem 2.17 since, from $\Psi^o(f,f) = 0$, it follows that $\max_{g \in \mathbb{D}} \Psi_o(g,f) \succcurlyeq 0$.
(Concerning the existence of this maximum cf. remark 2.12)

2.8. IWAMOTO [63] proved that under the assumptions (A2) and (A5): "For every $f \in \mathbb{D}$, there are an $x_o \in \mathbb{E}$, a number $n \in \mathbb{N}$, and an $\alpha > 0$ with $P(f)^n(x,x_o) \geqq \alpha$ for each $x \in \mathbb{E}$."
the assertions:

$$\Psi^o(g,f) \succcurlyeq 0 \Rightarrow g^g \geqq g^f$$

and, if for every $x \in \mathbb{E}$ being recurrent under g $\Psi^o(g,f) \succcurlyeq 0$, then

$$\psi^o(g,f)(x) > 0 \Rightarrow g^g(x) > g^f(x).$$

Since (A5), as a special case of the DOEBLIN-condition, does not only imply 1-recurrence but also the quasi-compactness of every $P(f)$, $f \in \mathbb{D}$, it is easy to recall the results mentioned above as special cases of theorem 2.17. The same can be said about a similar result given by KÜENLE [74].

2.9. We need not require the 1-recurrence of $P(g)$ to prove the following assertion:
Let the $(P,r)^*$-equations be solvable for certain $g,f \in \mathbb{D}$. Then the implication

$$\psi^{-1}(g,f) \geqq 0 \wedge \psi^o(g,f) \geqq 0 \Rightarrow g^g \geqq g^f$$

is valid. This result is due to ASO and KIMURA [1]. It is a special case of theorem 2.17 (i) since the assumption $\Psi^o(g,f) \succcurlyeq 0$ is essentially more general.

In the following let

$$\mathfrak{G}_o(f) := \{g \in \mathbb{D}:\ \psi^o(g,f) \succ 0\}$$

for every decision function whose $(P,r)^*$-equations are solvable.

Corollary 2.19. Let the $(P,r)^*$-equations of two decision functions $f,g \in \mathbb{D}$ be solvable. Furthermore, let $P(g)$ be 1-recurrent, $g \in \mathfrak{G}_o(f)$, and let the condition

$$\Psi^o(g,f)(x) = 0 \Rightarrow g(x) = f(x) \qquad (2.11)$$

be fulfilled for every $x \in \mathbb{E}$. Then it holds either that $g^g > g^f$ or $g^g = g^f$ and $v^g > v^f$.

Proof: As assumed, $\Psi^o(g,f) > 0$. Due to theorem 2.17 (i), it follows that $g^g \geq g^f$. This implies either $g^g > g^f$ or $g^g = g^f$ is valid. In the second case, condition C(x) must be fulfilled for every $x \in E$ as can be seen from theorem 2.17 (iii). Due to part (iv) of the same theorem we then have $v^g \geq v^f$. Now, because $\Psi^o(g,f) \neq 0$, there is a state x_0 satisfying $\phi^o(g,f)(x_0) > 0$. Therefore, $C^*(x_0)$ is valid, and it follows that $v^g(x_0) > v^f(x_0)$ due to part (v) of 2.17. Thus the proof is complete. ∎

Note: This result is due to VEINOTT [117] for the special case of E being finite.

Now we represent the policy improvement procedure already mentioned. For simplicity, we assume:

(v): For every $f \in \mathbb{D}$, the $(P,r)^*$-equations have a solution, and P(f) is recurrent.

Remark 2.10. The assumption (v) is fulfilled, in particular, if

1) E is finite,
2) the assumptions (A1) and (A2) are fulfilled,
3) the assumptions (A1), (A3), and (A4) are valid,

or if the condition:

4) (A5) is fulfilled

is valid (cf. remarks 2.6 - 2.8). With regard to the definition 1.8 and the **theorem 2.12 the condition:**

5) (A6): For every $f \in \mathbb{D}$, the assertion that P(f) is a strong Markov chain, will suffice for (v) to be fulfilled.

Now we assume (v). Let any decision function $f_0 \in \mathbb{D}$ be given. Then g^{f_0} and v^{f_0} are calculated from the $(P,r)^*$-equations. Two cases are possible: Either $\mathfrak{G}_0(f_0) \neq \emptyset$ or $\mathfrak{G}_0(f_0) = \emptyset$. In the first case a "new" decision function $f_1 \in \mathfrak{G}_0(f_0)$ is arbitrarily chosen. Due to part (i) of theorem 2.17 it follows immediately that the stationary policy f_1^∞ is at least as good as f_0^∞ with respect to the average reward criterion; that is, $g^{f_1} \geq g^{f_0}$. The procedure can then be continued with f_1 instead of f_0. - In the opposite case ($\mathfrak{G}_0(f_0) = \emptyset$) the procedure is stopped. Now we show that, in this case, an improvement of f_0 is not possible.

Theorem 2.20.

(i) Let $f,g \in \mathbb{D}$ be decision functions whose $(P,r)^*$-equations are solvable. If (P(g) is 1-recurrent and if $\Psi^o(g,f) \leqslant 0$, then $g^g \leq g^f$.
(ii) Let assumption (v) be fulfilled. If for a certain decision function f^* it holds that $\mathfrak{G}_0(f^*) = \emptyset$, then $(f^*)^\infty$ is average optimal on $\mathfrak{R}_S$.

Proof: Part (i) can be proved in a completely analogous manner to the corresponding part of theorem 2.17.
(ii) Since $\mathcal{B}_o(f^*) = \emptyset$, for every $g \in D$ we have $\Psi^o(g,f^*) \leqslant 0$. This implies $g^g \leq g^{f^*}$ in accordance with part (i).

Notes:
1. Theorem 2.20 is a generalization of similar results due to HOWARD [60] who assumed E to be finite.

2. Part (ii) of theorem 2.20 was mentioned by ASO and KIMURA [1] under assumptions which slightly differ form (A6); indeed, the proof given in [1] seems to be not correct.

Remarks:
2.11. If the assumption (v) is essentially strengthened, it can be proved that a decision function f^* satisfying $\mathcal{B}_o(f^*) = \emptyset$ is average optimal not only on $\mathfrak{R}_S$ but also on $\mathfrak{R}$ (e.g., in the finite case) and on $\mathfrak{R}_M$, respectively (cf. section 2.6).

2.12. Though the new iterated $f_1 \in \mathcal{B}_o(f_1)$ can, in general, be chosen in an arbitrary manner, we suggest doing so by taking into account the following points:
First: "good" decisions have to preserved, and second: one chooses a decision function f_1 maximizing the term $\Psi^o(f_1,f_o)$ in the lexicographical sense. Thus one applies the following selection procedure supposing that g^{f_o} and v^{f_o} are known:
For every state $x \in E$ one will find the sets

$$D_{-1}^{f_o}(x) := \{ d \in D(x) : \phi^{-1}(f_o)(x,d) = \max_{d' \in D(x)} \phi^{-1}(f_o)(x,d') \}$$

and

$$D_o^{f_o}(x) := \{ d \in D_{-1}^{f_o}(x) : \phi^o(f_o)(x,d) = \max_{d' \in D_{-1}^{f_o}(x)} \phi^{-1}(f_o)(x,d') \}$$

(both sets are non-empty since D(x) is finite). f_1 will be defined as follows:
If $f_o(x)$ is a "good" decision (this means that $f_o(x)$ is already a "good" decision), one defines $f_1(x) := f_o(x)$, otherwise, one defines $f_1(x) := d$ with $d \in D_o^{f_o}(x)$ arbitrarily chosen.

Now, the sets

$$\mathbb{D}_{-1}^{f_o} := \prod_{x \in E} D_{-1}^{f_o}(x) \text{ and } \mathbb{D}_o^{f_o} := \prod_{x \in E} D_o^{f_o}(x)$$

(both sets contain f_1) obviously satisfy the equations

$$\mathbb{D}_{-1}^{f_o} = \{ g^* \in \mathbb{D} : \phi^{-1}(g^*,f_o) = \max_{g \in \mathbb{D}} \phi^{-1}(g,f_o) \}$$

and

$$\mathbb{D}_o^{f_o} = \{ g^* \in \mathbb{D} : \Psi^o(g^*,f_o) = \max_{g \in \mathbb{D}} \Psi^o(g,f_o) \}.$$

This version of the policy improvement procedure is completely analogous to the "classical" procedure due to HOWARD [60].

2.13. The assertion of theorem 2.20 (i) cannot be reversed, i.e., $\Phi_o(f^*) = \emptyset$ need not hold, though f^* is an average reward decision function. The following example will show this fact.

Example 2.4. Let $E := \{1,2,3\}$ and $D(1) := D(3) := \{1\}$, $D(2) := \{1, 2\}$. There are only two different decision functions (which will be represented by vectors here):
$f := (1, 1, 1)^T$ and $g = (1, 2, 1)^T$. Further, let

$$P(f) = \begin{pmatrix} 1 & 0 & 0 \\ 0 & 1/2 & 1/2 \\ 0 & 1/3 & 1/3 \end{pmatrix}, \quad P(g) = \begin{pmatrix} 1 & 0 & 0 \\ 1/2 & 1/4 & 1/4 \\ 0 & 1/3 & 2/3 \end{pmatrix},$$

and

$$r(f) = r(g) = \begin{pmatrix} 16 \\ 10 \\ 20 \end{pmatrix}.$$

We obtain

$$P(f)^* = \begin{pmatrix} 1 & 0 & 0 \\ 0 & 2/5 & 3/5 \\ 0 & 2/5 & 3/5 \end{pmatrix} \text{ and } P(g)^* = \begin{pmatrix} 1 & 0 & 0 \\ 1 & 0 & 0 \\ 1 & 0 & 0 \end{pmatrix}.$$

Hence, due to (2.6)

$$g^f = g^g = \begin{pmatrix} 16 \\ 16 \\ 16 \end{pmatrix},$$

i.e., taking into account remark 2.11, both f^∞ and g^∞ are average optimal policies on $\mathcal{R}$. With respect to f the $(P,r)^*$-equations yield

$$v^f = \begin{pmatrix} 0 \\ -36/5 \\ 24/5 \end{pmatrix}.$$

Since D(1) and D(3) contain only one element, we obtain that $D_o^f(1) = D_o^f(3) = \{1\}$, and since g^f is constant, we have

$$D_{-1}^f(2) = \{1, 2\}.$$

Furthermore,

$$\psi^o(f)(2, 1) = 10 + (0 \ 1/2 \ 1/2) \begin{pmatrix} 0 \\ -36/5 \\ 24/5 \end{pmatrix} + 36/5 - 16 = 0,$$

but, on the contrary,

$$\psi^o(f)(2, 2) = 10 + (1/2 \ 1/4 \ 1/4) \begin{pmatrix} 0 \\ -36/5 \\ 24/5 \end{pmatrix} + 36/5 - 16 = 3/5$$

implying $D_o^f(2) = \{2\}$. Obviously, the average optimal decision function f satisfies: $\mathfrak{G}_0(f) = \{g\}$.

Note: Evidently, any decision function f^* satisfies $\mathfrak{G}_0(f^*) = \emptyset$, if and only if g^f and v^f are a solution of the functional equation system

$$P(f^*)g^{f^*} = \max_{f\in D} P(f)g^{f^*}$$

and

$$g^{f^*} + v^{f^*} = \max_{f\in D_{-1}^{f^*}} \{r(f) + P(f)v^{f^*}\}$$

with $D_{-1}^{f^*} = \prod_{x\in E} D_{-1}^{f^*}(x)$ (cf. remark 2.12).
Systems of functional equations of this or a similar type (so-called BELLMAN equations) are often used as a stationary policy in Markov Decision Theory. In view of example 2.4 it is clear that the existence problem of an average optimal policy is not identical with the solvability problem for such a BELLMAN equation. ∎

Theorem 2.20 justifies to stop the policy improvement procedure if a decision function f^* with $\mathfrak{G}(f^*) = \emptyset$ is found. This is exactly the case when the algorithm stops after a finite number of steps. Later on, we shall show this case to occur if E is finite. On the contrary, if E is countably infinite it is still an open question, which assumptions have to be required assuring such a decision function f^* to exist at all.
Unfortunately, there need not be a decision function f^* with $\mathfrak{G}_0(f^*) = \emptyset$ even if assumption (v) is fulfilled. We shall show this by an example whose parameters are chosen according to an idea of DERMAN [18].

Example 2.5. Let $E := \mathbb{N}_0$, $D(0) := \{1\}$, and $D(x) := \{1, 2\}$ for $x \in \mathbb{N}$. The transition law is given by

$$p(0,1,x) := \begin{cases} 0 & x = 0 \\ 2^{-x} & x \in \mathbb{N} \end{cases},$$

$$p(x,1,y) := \delta_{xy} \quad (x \in \mathbb{N}), \text{ and}$$

$$p(x,2,y) := \delta_{oy} \quad (x \in \mathbb{N}).$$

Let the pay-off structure be independent of the decisions, namely,

$$r(0) := r(0, 1) := 0 \text{ and}$$
$$r(x) := r(x, 1) := r(x, 2) := 1 - 2^{-x} \quad (x \in \mathbb{N}).$$

In order to show that the assumption (v) is fulfilled, we first remark that every decision function $f \in \mathbb{D}$ can be characterized one-to-one by a set J_f of indices in which the decision $d = 2$ is chosen:

$$J_f := \{x \in \mathbb{N}: f(x) = 2\}.$$

We have to distinguish three general cases:

1 st case: $J_f = \emptyset$. The corresponding decision function $f \equiv 1$ is denoted by f_0. We obtain

$$P(f_0) = \begin{pmatrix} 0 & 1/2 & 1/4 & 1/8 & \cdots \\ 0 & 1 & 0 & 0 & \cdots \\ 0 & 0 & 1 & 0 & \cdots \\ 0 & 0 & 0 & 1 & \cdots \\ \vdots & \vdots & \vdots & \vdots & \end{pmatrix}.$$

Evidently, $P(f_0)$ is idempotent, i.e., $P(f_0)^2 = P(f_0)$, hence, $P(f_0)^* = P(f_0)$. Each state $x \in \mathbb{N}$ forms a one-element (positive) recurrence class. The only transient state is 0, leading to $P(f_0)_{TT} = 0$. From this it is evident that $P(f_0)$ is 1-recurrent. ($P(f_0)$ is even a strong Markov chain.) A simple calculation shows that the $(P,r)^*$-equations have the solution:

$$g^{f_0}(x) = \begin{cases} 2/3 & x = 0 \\ 1 - 2^{-x} & x \in \mathbb{N} \end{cases} \qquad (2.12)$$

and

$$v^{f_0}(x) = \begin{cases} -2/3 & x = 0 \\ 0 & x \in \mathbb{N} \end{cases} .$$

2 nd case: $J_f = \mathbb{N}$. In this case we have $f_\infty := f = (1, 2, 2, \ldots)^T$. For the corresponding transition matrix it holds that

$$P(f_\infty) = \begin{pmatrix} 0 & 1/2 & 1/4 & 1/8 & \cdots \\ 1 & 0 & 0 & 0 & \cdots \\ 1 & 0 & 0 & 0 & \cdots \\ 1 & 0 & 0 & 0 & \cdots \\ \vdots & \vdots & \vdots & \vdots & \end{pmatrix}, \quad P(f_\infty)^2 = \begin{pmatrix} 1 & 0 & 0 & 0 & \cdots \\ 0 & 1/2 & 1/4 & 1/8 & \cdots \\ 0 & 1/2 & 1/4 & 1/8 & \cdots \\ 0 & 1/2 & 1/4 & 1/8 & \cdots \\ \vdots & \vdots & \vdots & \vdots & \end{pmatrix}.$$

By induction it can be shown that the equations

$$P(f_\infty)^{2n+1} = P(f_\infty) \text{ and } P(f_\infty)^{2n} = P(f_\infty)^2$$

are valid for all $n \in \mathbb{N}$. From this follows that

$$P(f_\infty)^* = \begin{pmatrix} 1/2 & 1/4 & 1/8 & \cdots \\ 1/2 & 1/4 & 1/8 & \cdots \\ 1/2 & 1/4 & 1/8 & \cdots \\ \vdots & \vdots & \vdots & \end{pmatrix},$$

i.e., $P(f_\infty)$ is non-dissipative and, henceforth, 1-recurrent since there are no transient states. ($P(f_\infty)$ is even quasi-compact, because $P(f_\infty)(x,A) \geq \frac{1}{2}$ for the set $A := \{0, 1\}$ and all $x \in E$ – the DOEBLIN condition.) Now, we show that the $(P,r)^*$-equations have a solution. Since E forms one class of recurrent states g^{f_∞} must be of the form $g^{f_\infty} = g\,\mathbf{1}$ with $g \in \mathbb{R}$. The solution of (2.8') and (2.9) yields

$$g^{f_\infty} = \frac{1}{3}\,\mathbf{1} \qquad (2.13)$$

and

$$v^{f_\infty} = \begin{cases} -1/6 & x = 0 \\ 2^{-x}(2^{x-1}-1) & x \in \mathbb{N} \end{cases} . \qquad (2.14)$$

<u>3 rd case:</u> $\emptyset \neq J_f \overset{\subset}{\neq} \mathbb{N}$ turns out to be somewhat more complicated. J_f can be written as $J_f = \{i_1, i_2, \ldots\}$ with $i_1 < i_2 < \ldots$ Let $T := J_f \cup \{0\}$ and $R := E\backslash T$. Then R is enumerated in an analogous manner: $R = \{j_1, j_2, \ldots\}$ with $j_1 < j_2 < \ldots$ If we set $p_x := 2^{-x}$, then, after a suitable ordering the states, one has

$$P(f) = \begin{array}{c} \\ \vdots \\ j_3 \\ j_2 \\ j_1 \\ \\ \\ i_1 \\ i_2 \\ i_3 \\ \vdots \end{array} \left(\begin{array}{cccc|ccccc} \cdots & j_3 & j_2 & j_1 & 0 & i_1 & i_2 & i_3 & \cdots \\ & \vdots & \vdots & \vdots & & & & & \\ \cdots & 1 & 0 & 0 & 0 & 0 & 0 & 0 & \cdots \\ \cdots & 0 & 1 & 0 & 0 & 0 & 0 & 0 & \cdots \\ \cdots & 0 & 0 & 1 & 0 & 0 & 0 & 0 & \cdots \\ \hline \cdots & p_{j_3} & p_{j_2} & p_{j_1} & 0 & p_{i_1} & p_{i_2} & p_{i_3} & \cdots \\ \cdots & 0 & 0 & 0 & 1 & 0 & 0 & 0 & \cdots \\ \cdots & 0 & 0 & 0 & 1 & 0 & 0 & 0 & \cdots \\ \cdots & 0 & 0 & 0 & 1 & 0 & 0 & 0 & \cdots \\ & \vdots & \vdots & \vdots & \vdots & \vdots & \vdots & \vdots & \end{array} \right) .$$

By means of induction one obtains the equations

$$(P(f)_{TT})^{2n+1} = \alpha^n\, P(f)_{TT} \text{ and } (P(f)_{TT})^{2n} = \alpha^{n-1}(P(f)_{TT})^2$$

$(n \in \mathbb{N})$. Since $J_f \overset{\subset}{\neq} \mathbb{N}$ we have $\alpha < 1$. Thus for $n \in \mathbb{N}$

$$\sum_{m=0}^{2n} (P(f)_{TT})^m\, \mathbf{1}_T = \mathbf{1}_T + \frac{1-\alpha^n}{1-\alpha} \begin{pmatrix} 2 \\ 1+\alpha \\ 1+\alpha \\ \vdots \end{pmatrix} ,$$

this implies

$$\sum_{m=0}^{\infty} (P(f)_{TT})^m\, \mathbf{1}_T = \mathbf{1}_T + \frac{1}{1-\alpha} \begin{pmatrix} 2 \\ 1+\alpha \\ 1+\alpha \\ \vdots \end{pmatrix} .$$

This means, however, that P(f) is a strong Markov chain and, in particular, 1-recurrent. The solution of $(P,r)^*$-equations with respect to f can be found as follows:

The equation $g^f = P(f)\, g^f$ yields

$$g(0) = g(i_1) = \ldots = g(i_n) = \ldots = \frac{1}{1-\alpha} \sum_n p_{j_n}\, r(j_n). \qquad (*)$$

From $P(f)^* v^f = 0$ it follows that $v^f(x) = 0$ for $x \in R$. Hence the equation

$$g^f + v^f = r(f) + P(f)v^f,$$

which can be decomposed into the partial equations

$$g^f(x) = r(x), \quad (x \in R),$$

$$v^f(0) = \sum_n p_{i_n} v(i_n) - g^f(0), \qquad (\ddagger)$$

and

$$v^f(i_n) = r(i_n) + v^f(0) - g^f(0) \qquad (\ddagger *)$$

has a unique solution, which can be calculated by comparing (*), (‡), and (‡*). This solution is

$$g^f(x) = \begin{cases} r(x) & x \in R \\ \frac{1}{1-\alpha} \sum_n p_{j_n} r(j_n) & x \in T \end{cases} \qquad (2.15)$$

and

$$v^f(x) = \begin{cases} 0 & x \in R \\ (1-\alpha)^{-1} \sum_n p_{i_n} r(i_n) - (1+\alpha)(1-\alpha)^{-2} \sum_n p_{j_n} r(j_n) & x = 0 \\ r(x) + (1-\alpha)^{-1} \sum_n p_{i_n} r(i_n) - 2(1-\alpha)^{-2} \sum_n p_{j_n} r(j_n) & x \in J_f . \end{cases} \qquad (2.16)$$

Thus it was shown that, for every decision function $f \in \mathbb{D}$, $P(f)$ is 1-recurrent and the $(P,r)^*$-equations have a solution, i.e., the assumption (v) is fulfilled.

It remains to show that there is no decision function $f \in \mathbb{D}$ satisfying $\mathfrak{G}_o(f) = \emptyset$. For this purpose we consider the decision functions defined by

$$f_n(x) := \begin{cases} 1 & x = 0 \text{ or } x \geq n + 1 \\ 2 & x = 1,2,\dots,n \end{cases},$$

$n \in \mathbb{N}$. It holds that $J_{f_n} = \{1,\dots,n\}$, and, due to (2.15), it follows that

$$g^{f_n}(0) = 1 - \frac{2^{-n}}{3} .$$

Hence $\sup_{n \in \mathbb{N}} g^{f_n}(0) = 1$. On the other hand, from (2.12), (2.13), and (2.15) (since always $r(j_n) < 1$) we can easily see that a decision function $f \in \mathbb{D}$ with $g^f(0) = 1$ cannot exist. Consequently, there does not exist any decision function f^* fulfilling the condition $\mathfrak{G}_o(f^*) = \emptyset$ at all. (The values $g^f(0)$, $f \in \mathbb{D}$, were already calculated by DERMAN [18], who used the model constructed above for proving the existence of a so-called "randomized" decision function f' satisfying $g^{f'}(o) = 1$).

Note: Using assumptions which are essentially more restrictive than (v), DERMAN [18], TAYLOR [110], and KUSHNER [75] proved that there is a decision function f^* which satisfies $\mathfrak{G}_0(f^*) = \emptyset$ (cf. page 83).

Example 2.5 shows that the policy improvement procedure need not stop after finitely many steps if the state space E is infinite, at least if there is no decision function f^* with $\mathfrak{G}_0(f^*) = \emptyset$. But even if such a function does exist, we cannot exclude that it cannot be found after finitely many steps. Concerning the sequence of decision functions obtained by a repeated application of the policy improvement procedure the following statement can be asserted.

Lemma 2.21. Let assumption (v) be fulfilled and let $(g_0, g_1, \ldots, g_n, \ldots)$ be a finite or infinite sequence of decision functions satisfying $g_{n+1} \in \mathfrak{G}_0(g_n)$ for each n. Then

(i) (g^{g_n}) is a non-decreasing sequence.

(ii) If, for each n, the additional assumption:

$\Psi^0(g_{n+1}, g_n)(x) = 0 \Rightarrow g_{n+1}(x) = g_n(x), \; x \in E,$

is fulfilled, then, furthermore, the implication

$$g_m = g_n \Rightarrow m = n$$

is valid.

Proof: Part (i) can directly be concluded from corollary 2.18 (i). Concerning part (ii) we assume **in contrast to the assertion,** that for certain n, k ∈ $\mathbb{N}$,

$$g_{n+k} = g_n,$$

then by (i)

$$g^{g_n} = g^{g_{n+1}} = \ldots = g^{g_{n+k}} .$$

Due to corollary 2.19 we would obtain that

$$v^{g_n} < v^{g_{n+1}} < v^{g_{n+k}},$$

this is a contradiction to the assertion.

Part (ii) of lemma 2.21 contains the noteworthy fact that the application of the HOWARD algorithm (in the sense of remark 2.12) involves that cycles are excluded, i.e., no decision function which has once been found by the algorithm can be found for a second time in the proceeding iteration. This is non-trivial as far as the sequence (g^{g_n}) although being non-decreasing need not be strictly increasing. (E.g., consider the functions f and g from example 2.4: It holds that $g \in \mathfrak{G}_0(f)$ and $g^g = g^f$.)

Since the algorithm does not generate cycles we obtain

Corollary 2.22. If the state space E is finite, the HOWARD-algorithm (in the sense of remark 2.12) can always be applied and stops after finitely many steps. The last iterated f* satisfies $\mathbb{G}_0(f^*) = \emptyset$.

Proof: It suffices to note that $\mathbb{D}$ is finite. ∎

Note: This statement is due to HOWARD [60] who did not prove the fact that any cycles could not occur in the improvement procedure. The latter was shown by BLACKWELL [9] applying a technique being efficient only in the finite case.

Lemma 2.21 includes another interesting aspect connected with the following convergence notion introduced with respect to the set $\mathbb{D}$:
A sequence $(g_n)_{n\in\mathbb{N}}$ from $\mathbb{D}$ is said to be convergent to a function $g \in \mathbb{D}$ (in symbols: $\lim'_{n\to\infty} g_n = g$) if, for each $x \in E$, there is an $n(x) \in \mathbb{N}$ with $g_n(x) = g(x)$ for every $n \geq n(x)$. $\mathbb{D}$ is a compact set in the topology induced by this convergence.

Lemma 2.23. Let assumption (v) be fulfilled and let $(g_0, g_1, \ldots)$ be an infinite sequence of decision functions with $g_{n+1} \in \mathbb{G}_0(g_n)$ for every n. Then there is a bounded function g^∞: $E \to R$, this satisfies

$$\lim_{n\to\infty} g^{g_n} = g^\infty$$

(component-wise). Furthermore, there is a decision function $g_\infty \in \mathbb{D}$ such that the equation

$$\lim_{k\to\infty} g_{n_k} = g_\infty$$

is valid for a certain subsequence (g_{n_k}) of (g_n).

Proof: The first statement results directly from the monotony of the sequence (g^{g_n}) (cf. lemma 2.21) and its boundedness by k. The second statement is only a re-formulation of the compactness of $\mathbb{D}$. ∎

For the function g^∞ quoted in lemma 2.23

$$g^\infty \leq \sup_{f\in\mathbb{D}} g^f.$$

The authors do not know whether the case "<" can occur under the assumption (v). If the assumptions are suitably sharpened, this case cannot occur. For instance, DERMAN [18] and KUSHNER have shown that equality always holds (cf. theorem 2.24). Furthermore, lemma 2.23 involves the question whether the equation $g^\infty = g^{g_\infty}$ is valid. If this were the case the following useful consequence would result: If, during the iteration procedure, the sequence (g_n) or a certain subsequence of it converges to a decision function g_∞, then g_∞ is "better" than each element of the sequence (g_n). It is justified to continue the procedure with g_∞ as "start function". Unfortunately, the assumption (v) is not sufficient for the validity of the equation $g^\infty = g^{g^\infty}$ as the following example will show.

Example 2.6. The model and all notations are chosen as in example 2.5. We intend to investigate the sets $\mathfrak{G}_0(f)$ for several start functions $f \in \mathbb{D}$. For this purpose we note that

$$\psi^{-1}(f)(x,1) = 0,$$

$$\psi^{-1}(f)(x,2) = g^f(0) - g^f(x),$$

$$\psi^0(f)\ (x,1) = r(x) - g^f(x), \text{ and}$$

$$\psi^0(f)\ (x,2) = r(x) + v^f(0) - v^f(0) - g^f(x)$$

for every start function $f \in \mathbb{D}$ and all $x \in \mathbb{N}$, i.e., that is,

$$D^f_{-1}(x) = \begin{cases} \{1\} & \text{if } g^f(0) - g^f(x) < 0 \\ \{1, 2\} & \text{if } g^f(0) - g^f(x) = 0 \\ \{2\} & \text{if } g^f(0) - g^f(x) > 0 \end{cases}$$

and, if $D^f_{-1}(x) = \{1, 2\}$,

$$D^f_0(x) = \begin{cases} \{1\} & \text{if } v^f(0) - v^f(x) < 0 \\ \{1,2\} & \text{if } v^f(0) - v^f(x) = 0 \\ \{2\} & \text{if } v^f(0) - v^f(x) > 0 \end{cases} .$$

1) Start with $f_\infty = (1, 2, 3, \ldots)^T$: Since $g^f = \frac{1}{3}\,1$ only $\psi^0(f_\infty)$ has to be considered. Now, due to (2.14), for $x \in \mathbb{N}$, we have

$$v^{f_\infty}(0) - v^{f_\infty}(x) = -\frac{1}{6} - 2^{-x}(2^{x-1} - 1) < 0.$$

Hence $D_0^{f_\infty}(x) = \{1\}$ for each x. Thus $\mathfrak{G}_0(f_\infty) = \{f_0\}$.

2) Start with $f_0 = (1, 1, 1, \ldots)^T$: In connection with (2.12) we have

$$g^{f_0}(0) - g^{f_0}(x) = \frac{2}{3} - (1 - 2^{-x}) = 2^{-x} - \frac{1}{3}, \text{ i.e.,}$$

$$g^{f_0}(0) - g^{f_0}(x) = \begin{cases} > 0 & \text{if } x = 1 \\ < 0 & \text{if } x \geq 2 \end{cases},$$

which implies that $D_0^{f_0}(1) = \{2\}$, $D_0^{f_0}(x) = \{1\}$ for $x \geq 2$. Therefore $\mathfrak{G}_0(f_0) = \{f_1\}$.

3) Start with f_n, $n \in \mathbb{N}$: We obtain that $J_{f_n} = \{x \in \mathbb{N}: f_n(x) = 2\} = \{1, 2, \ldots, n\}$ for every f_n. Example 2.5 showed that $g^{f_n}(0) = 1 - \frac{1}{3} \cdot 2^{-n}$. Due to (2.15) it follows that

$$g^{f_n}(0) - g^{f_n}(x) = \begin{cases} 0 & \text{if } x \in J_{f_n} \\ 1 - \frac{1}{3} \cdot 2^{-n} - (1 - 2^{-x}) = 2^{-x} - \frac{1}{3} \cdot 2^{-n} & \text{otherwise.} \end{cases}$$

Therefore, $D_{-1}^{f_n}(x) = \{1, 2\}$ for $x \in J_{f_n}$.
Since for $x \notin J_{f_n}$ it holds that

$$2^{-x} - \frac{1}{3} \cdot 2^{-n} = \begin{cases} > 0 & \text{if } x = n+1 \\ < 0 & \text{if } x \geq n+2 \end{cases}$$

we obtain

$$D_{-1}^{f_n}(n+1) = \{2\} \text{ and } D_{-1}^{f_n}(x) = \{1\}$$

for $x \geq n + 2$. It remains to investigate $D_0^{f_n}(x)$ for $x \in J_{f_n}$. This has to be done by considering the difference $v^{f_n}(0) - v^{f_n}(x)$. Due to (2.16), for $x \in J_{f_n}$, the inequality

$$v^{f_n}(0) - v^{f_n}(x) = (1 - \alpha)^{-1} \sum_m p_{j_m} r(j_m) - r(x) = g^{f_n}(0) - r(x)$$

$$= 2^{-x} - \frac{1}{3} \cdot 2^{-n} \geq 2^{-n} - \frac{1}{3} \cdot 2^{-n} > 0$$

is valid, this implies $D_0^{f_n}(x) = \{2\}$ for $x \in J_{f_n}$. It results that $\mathfrak{G}_0(f_n) = \{f_{n+1}\}$.

From the above considerations we can conclude the following:
If the iteration starts with f_0, the resulting sequence $(g_n)_{n \in \mathbb{N}_0}$ of decision functions is the very $(f_n)_{n \in \mathbb{N}_0}$, where each of the g_n (except for g_0) is calculated from the preceding one. Roughly speeking, to do one policy improvement step just means to choose decision 2 one time more that it has been the case for the preceding decision function. Obviously,

$$\lim_{n \to \infty} f_n = f_\infty = g_\infty ,$$

and (as it was already shown in the example 2.5)

$$\lim_{n \to \infty} g^{f_n}(0) = g^\infty(0) = 1.$$

We have, however, $g^{f_\infty}(0) = \frac{1}{3}$. Thus in the state 0, the limit policy $(f_\infty)^\infty$ is "worse" than each iterated $(f_n)^\infty$, $n \in \mathbb{N}_0$. This does not only show that, in general, the equation $g^{g^\infty} = g^\infty$ need not be valid, but also that the continuation of the improvement procedure with the limit decision function g_∞ can be rather disadvantageous if $g^{g^\infty} \neq g^\infty$. In our example this would mean to start the whole iteration once again from the very beginning. ▮

At the end of this subsection we shall cite two results concerning the convergence of the sequence (g_n) of decision functions, which are improved step by step by using the HOWARD algorithm (or a subsequence of it), to an optimal decision function. For this purpose we make use of the following assumptions:

(A7): For each $y \in E$

$$\inf_{\substack{f \in D \\ x \in E}} P(f)^*(x,y) > 0.$$

(A8): $P(f)$ is aperiodic for every $f \in D$.

Theorem 2.24.

(i) (DERMAN [18]) If the assumptions (A1), (A2), and (A6) are fulfilled and both the average rewards g^f and the functions v^f are uniformly bounded in f, then there is a decision function g^* satisfying $\mathcal{B}_0(g^*) = \emptyset$.

If, in addition, assumption (A7) is fulfilled, then, whatever the choice of g_0, the sequence of decision functions generated by the HOWARD-algorithm (in the sense of remark 2.12) converges to a decision function g^* with $\mathcal{B}_0(g^*) = \emptyset$. The stationary policy $(g^*)^\infty$ is average optimal on $\mathcal{R}$.

(ii) (KUSHNER [75]) Under the assumptions (A1), (A3), (A4), and (A8) it holds that

$$\lim_{n \to \infty} g^{g_n} = g^*$$

for the sequence $(g_n)_{n \in N_0}$ generated by the HOWARD algorithm. A certain (g_{n_k}) converges to a decision function g^* which satisfies

$$g^{g^*} + v^{g^*} = \max_{g \in D} \{r(g) + P(g) v^{g^*}\}.$$

The policy $(g^*)^\infty$ is average optimal.

EXISTENCE OF AVERAGE OPTIMAL POLICIES

Now we shall give two results concerning the existence of average reward optimal policies. These results are motivated by the situation arising when the HOWARD algorithm stops, although the applicability of this algorithm is not supposed.

Theorem 2.25. (DERMAN [18])[1]) If there exist a constant $g \in \mathbb{R}$ and a bounded real function v on E which satisfy the functional equation

$$g\,1 + v = \max_{f \in D} \{r(f) + P(f)v\}, \qquad (2.17)$$

then there is an average reward optimal stationary policy. For each decision function f^* with

$$g\,1 + v = r(f^*) + P(f^*)v$$

the equation

[1]) See also DERMAN and LIEBERMAN [21].

$$g\,\mathbf{1} = g^{f^*} = g^*$$

is valid. The stationary policy $(f^*)^\infty$ is average optimal on $\mathcal{R}$.

Notes:

1. Let us remark that assumption (v) is not required in this theorem.

2. If (v) is fulfilled the situation described by theorem 2.25 is under consideration, if and only if a decision function f^* can be found with $\mathcal{C}_0(f^*) = \emptyset$ and $g^{f^*} = g^{f^*}\mathbf{1}$ (g^{f^*} is a suitable constant). With regard to the example 2.5 it is clear that (v) is not sufficient for the latter condition to be fulfilled.

3. Theorem 2.24 provides sufficient conditions for the system (2.17) having a bounded solution. As already mentioned these conditions are sharper than (v).
Further sufficient conditions were given by TAYLOR [110], ROSS [96], [97], GUBENKO and STATLAND [39], TIJMS [112], and YAMADA [125]. Without going into details we remark that the assumptions used there imply the validity of (v) if they are applied to the countable state case.

4. Using an assumption of the so-called Ljapunov-type (cf. KUSHNER [75]) HORDIJK and SLADKY [58] showed that the system (2.17) need not necessarily be bounded (see also the final part of 2.6).

The following sharpened version of corollary 2.22 will be used in chapter 3.

Theorem 2.26. (DERMAN [19]) If the state space E is finite, then there are vectors q and v which satisfy the system of equations

$$g = \max_{f \in D} P(f) g \tag{2.18}$$

and

$$g + v = \max_{f \in D^*_{-1}} \{ r(f) + P(f) v \}, \tag{2.19}$$

where $D^*_{-1} := \{ g^* \in D:\ g = P(g^*) g \}$. For each decision function $f^* \in D^*_0$ with

$$D^*_0 = \{ f^* \in D^*_{-1}:\ g + v = r(f^*) + P(f^*) v \}$$

it holds that

$$g = g^{f^*} = g^*,$$

whereas the stationary policy $(f^*)^\infty$ is average optimal.

Remark 2.14. It is possible to make assertions concerning the existence of stationary policies being average optimal on $\mathcal{R}$ or certain subsets of $\mathcal{R}$ without referring to the functional equations (2.17) and (2.18) together with (2.19), respectively. Several authors choose another approach (e.g., BATHER [2], FEINBERG [30], HORDIJK [54], [55], VISKOV and SHIRJAJEV [118], and WIJNGAARD [121], [124]. They obtained general existence theorems. However, the question, how to examine a given policy concerning the problem whether it is average optimal or not, was not answered.

Note: To a certain extent it is possible to reformulate the assertions made above for the continuous-time case. KAKUMANU [67], [68], BATHER [3], and DOSHI [27] investigated this case. Though restricting themselves to "singlé-chains", among other results, they showed that a functional equation similar to (2.17) guarantees the existence of stationary average optimal policy. Furthermore, KAKUMANU and DOSHI proved that the "single-chain" version of the HOWARD algorithm can be applied under suitable conditions.

2.5. OPTIMIZATION BASED ON SENSITIVE CRITERIA

The present section is mainly based on the investigations which were carried out by DIETZ [24] concerning sensitive optimality criteria.

Above all a generalized version of the HOWARD algorithm for the stepwise improvement of stationary policies with respect to these criteria is developed. This can be applied not only if the state space is finite and, but also under the supposition of the DOEBLIN condition, if it is countably infinite. This algorithm essentially is based on a Laurent expansion of the discounted total reward expected under a stationary policy. The "ordinary" HOWARD algorithm given in the preceding section will turn out to be a special case of the procedure introduced here.

Using a somewhat sharper assumption than the DOEBLIN condition we succeed in characterizing the sensitive discount optimality of a policy within the class $\mathfrak{R}_M$ of Markov policies. We conclude this section with a collection of results concerning the existence of sensitive discount optimal policies which have been published by other authors.

LAURENT EXPANSION OF THE DISCOUNTED TOTAL REWARD

Let $\beta \in (0,1)$ be a discount factor. In order to simplify the following considerations we define $\rho := \beta^{-1}(1-\beta)$, the so-called "interest rate".

Let $f \in D$ be an arbitrary decision function. If the transition operator $P(f)$ is quasi-compact (i.e., $P(f)$ satisfies the DOEBLIN-condition), we set

$$H(f) := (I - (P(f) - P(f)^*)^{-1} - P(f)^*.$$

Theorem 2.27: Let $f \in \mathbb{D}$ and let $P(f)$ be quasi-compact. Then for every $\rho \in (0, \|H(f)\|^{-1})$ the expansion

$$V_\beta^f = (1+\rho) \sum_{n=-1}^{\infty} v_n^f \rho^n \tag{2.20}$$

is valid, where

$$v_{-1}^f = P(f)^* r(f) \tag{2.21}$$

and

$$v_n^f = (-1)^n H(f)^{n+1} r(f) \quad (n \in \mathbb{N}_0). \tag{2.22}$$

The series (2.20) absolutely converges in $B(E)$.

Proof: Due to (2.5) it holds that

$$V_\beta^f = (I - \beta P(f))^{-1} r(f)$$

and therefore

$$\begin{aligned} V_\beta^f &= ((1+\rho)^{-1}((1+\rho)I - P(f)))^{-1} r(f) \\ &= -(1+\rho)((P(f) - I) - \rho I)^{-1} r(f) \\ &= -(1+\rho) R(\rho, L) r(f) \end{aligned}$$

with $L := P(f) - I$ and $R(\rho, L) = (L - \rho I)^{-1}$.
In view of theorem 1.18 the equation

$$R(\rho, L) = -\rho^{-1} P(f)^* + \sum_{n=0}^{\infty} (-1)^{n+1} \rho^n H(f)^{n+1}$$

$(\rho \in (0, \|H(f)\|^{-1}))$ holds in $\mathfrak{L}(B(E))$. From this it results that

$$\begin{aligned} V_\beta^f &= -(1+\rho)(-\rho^{-1} P(f)^* + \sum_{n=0}^{\infty} (-1)^{n+1} \rho^n H(f)^{n+1}) r(f) \\ &= (1+\rho)(\rho^{-1} P(f)^* r(f) + \sum_{n=0}^{\infty} (-1)^n \rho^n H(f)^{n+1} r(f)) \\ &= (1+\rho) \sum_{n=-1}^{\infty} \rho^n v_n f. \end{aligned}$$

Since $\rho\|H(f)\| =: \alpha < 1$, due to the assumption, the absolute convergence of this series follows from the inequality

$$\begin{aligned} \|\rho^n v_n^f\| &= \|(-1)^n \rho^n H(f)^{n+1} r(f)\| \leq (\rho\|H(f)\|)^n \|H(f) r(f)\| \\ &= \alpha^n \|H(f) r(f)\| \end{aligned}$$

for $n \geq 0$. ∎

Notes:

1. Theorem 2.27 is due to DIETZ [24]; the expansion (2.20) was also deduced by KADOTA [66] independently of [24].

2. If $P(f)$ is quasi-compact and has only one class of positive-recurrent states the expansion (2.20) was derived by HORDIJK and SLADKY [58].

Under weaker conditions the same authors proved a partial Laurent expansion for the "single-chain" case in the form

$$V^f_\beta = (1+\rho)\sum_{n=-1}^{M} \rho^n v^f + O(\rho^M).$$

If one considers only the first two terms of the series (2.20) and applies β instead of ρ, then one receives the asymptotic equation

$$V^f_\beta = (1-\beta)^{-1}\, v^f_{-1} + v^f_0 + o(1)$$

for $\beta \uparrow 1$.

This representation was derived by BLACKWELL [9] for the finite state case and by KUSHNER [75] for the case of countable state space, where he required the assumptions (A1), (A4), and (A8) from section 2.4.

3. The expansion (2.20) is always valid if the state space is finite since then every operator $P(f)$, $f \in D$, is quasi-compact. The validity of (2.20) in this special case was shown by MILLER and VEINOTT [84].

Remark 2.15. The coefficient v^f_{-1} of the expansion (2.20) coincides with the average return g^f of f as can be seen by lemma 2.8. If $P(f)$ is quasi-compact, then, due to corollary 2.13, the $(P,r)^*$-equations possess a solution. By consideration of the identities

$$\begin{aligned} r(f) + P(f)v^f_0 &= r(f) + P(f)H(f)r(f) \\ &= r(f) + (H(f) - I - P(f)^*)r(f) = \\ &= g^f + v^f_0 \end{aligned}$$

and

$$P(f)^* v^f_0 = P(f)^* H(f) r(f) = 0,$$

which can be obtained by theorem 1.17, the equality $v^f = v^f_0$ follows. In addition to remark 2.3, the vectors g^f and v^f can be interpreted as follows: The rate of increase of the β-discounted total reward of a stationary policy f^∞, $f \in D$, whereas β tends to 1, depends, first of all, on the average expected reward g^f and secondarily on the "weight" vector v^f.

By theorem 2.27 it is possible to compare two stationary policies with respect to sensitive discount optimality criteria. To do this we consider the (finite or infinite) matrices

$$\mathfrak{v}(f) := (v^f_{-1}, v^f_0, v^f_1, \ldots) \text{ and } \mathfrak{v}_n(f) := (v^f_{-1}, v^f_0, \ldots, v^f_n),$$

$n \in \mathbb{N}_{-1}$, for each $f \in D$. Their rows corresponding to each $x \in E$ will be denoted by

$$\mathfrak{v}(f)(x) := (v^f_{-1}, v^f_0, \ldots, v^f_n, \ldots)(x) := (v^f_{-1}(x), v^f_0(x), \ldots, v^f_n(x), \ldots)$$

and

$$\mathfrak{v}(f)(x) := (v^f_{-1}, v^f_0, \ldots, v^f_n)(x) := (v^f_{-1}(x), v^f_0(x), \ldots, v^f_n(x)).$$

(We set $v^f_{-2} := 0$.)

Lemma 2.28. Let $f, g \in \mathbb{D}$, and let $P(f)$ and $P(g)$ be quasi-compact. Then, for every $N \in \mathbb{N}_{-1}$, the following equivalence relations hold:

$$\lim_{\beta\to 1}\left(\frac{1-\beta}{\beta}\right)^{-N}(V_\beta^g - V_\beta^f)(x) \geq 0 \Leftrightarrow \mathfrak{v}_N(g)(x) \succcurlyeq \mathfrak{v}_N(f)(x)$$

and

$$\lim_{\beta\to 1}\left(\frac{1-\beta}{\beta}\right)^{-N}(V_\beta^g - V_\beta^f)(x) > 0 \Leftrightarrow \mathfrak{v}_N(g)(x) \succ \mathfrak{v}_N(f)(x),$$

$x \in \mathbb{E}$.

Proof: Due to theorem 2.27 it is justified to replace "lim inf" by "lim". From the same theorem it follows that

$$V_\beta^f = (1+\rho)\sum_{n=-1}^{\infty} \rho^n \mathfrak{v}_n^f = \sum_{n=-1}^{\infty} \rho^n(\mathfrak{v}_n^f + \mathfrak{v}_{n-1}^f).$$

Thus

$$\rho^{-N}(V_\beta^g - V_\beta^f) = \left(\frac{1-\beta}{\beta}\right)^{-N}(V_\beta^g - V_\beta^f) =$$

$$= \sum_{n=-1}^{N} \rho^{n-N}((\mathfrak{v}_n^g + \mathfrak{v}_{n-1}^{g-1}) - (\mathfrak{v}_n^f + \mathfrak{v}_{n-1}^f)) + O(1)$$

for $\rho \to 0$; this implies

$$\lim_{\beta\to 1}\left(\frac{1-\beta}{\beta}\right)^{-N}(V_\beta^g - V_\beta^f)(x) = \lim_{\rho\to 0}\sum_{n=-1}^{N} \rho^{n-N}(\mathfrak{v}_n^g - \mathfrak{v}_n^f + \mathfrak{v}_{n-1}^g - \mathfrak{v}_{n-1}^f)(x).$$

This term is non-negative, if and only if

$$(\mathfrak{v}_{-1}^g - \mathfrak{v}_1^f, \mathfrak{v}_0^g - \mathfrak{v}_0^f + \mathfrak{v}_{-1}^g - \mathfrak{v}_{-1}^f, \ldots, \mathfrak{v}_N^g - \mathfrak{v}_N^f + \mathfrak{v}_{N-1}^g - \mathfrak{v}_{N-1}^f)(x) \succcurlyeq 0;$$

this is equivalent to

$$\mathfrak{v}_N(g)(x) - \mathfrak{v}_N(f)(x) \succcurlyeq 0.$$

Analogous considerations lead to the assertion concerning "$\succ$".

Corollary 2.29. Let $N \in \mathbb{N}_{-1}$ and let $P(f)$ be quasi-compact for each decision function $f \in \mathbb{D}$. Any policy f^∞, $f \in \mathbb{D}$, is N-discounted optimal on $\mathfrak{R}_S$, if and only if for every $g \in \mathbb{D}$ the inequality

$$\mathfrak{v}_N(f) \succcurlyeq \mathfrak{v}_N(g)$$

is valid.

Proof: Following the definition 2.16 and theorem 2.27 f^∞ is N-discount optimal on $\mathfrak{R}_S$, if and only if

$$\liminf_{\beta\to 1}\left(\frac{1-\beta}{\beta}\right)^{-N}(V_\beta^f - V_\beta^g) = \lim_{\beta\to 1}\left(\frac{1-\beta}{\beta}\right)^{-N}(V_\beta^f - V_\beta^g) \geq 0$$

for all $g \in \mathbb{D}$. This is, however, equivalent to $\mathfrak{v}_n(f) \succcurlyeq \mathfrak{v}_N(g)$ by lemma 2.28. ∎

Notes:

1. In the following we denote the next assumption by
(V*): P(f) is quasi-compact for each $f \in \mathbb{D}$.
Now, in particular, by corollary 2.29 it is obvious that under (V*) a policy f^∞ is (-1)-optimal on $\mathfrak{R}_S$, if and only if it is average optimal on $\mathfrak{R}_S$. This assertion - with $\mathfrak{R}_S$ replaced by $\mathfrak{R}_M$ - remains valid if the assumptions are suitably sharpened as will be shown at the end of this section.

If the state space E is finite, an even more stringent assertion is true. Due to theorem 2.26 it can be concluded that the average optimality on $\mathfrak{R}_S$ coincides with the average optimality on $\mathfrak{R}$. The same can be asserted with respect to the N-optimality since BLACKWELL [9] proved the existence of a stationary policy which is N-optimal on $\mathfrak{R}$ for each $N \in \mathbb{N}_{-1}$. Therefore, the notions "(-1)-optimality" and "average optimality" coincide in the finite case. If the state space is countably infinite, this need not be true under general assumptions as can be seen by an example given by FLYNN [33].

2. If the assumption (V*) is suitably sharpened, 0-optimality on $\mathfrak{R}_S$ and 0-optimality on $\mathfrak{R}$ are equivalent. This was shown by KADOTA [66].

3. Interesting considerations concerning the relationship between sensitive discount criteria and those which are based on an iterated averaging procedure ("sensitive averaging criteria") were made by SLADKY [105], [106] (finite case) and by HORDIJK and SLADKY [58] and FLYNN [34](countable case) (cf. also VEINOTT [117], LIPPMAN [77], and DENARDO and MILLER [17]).

A POLICY IMPROVEMENT PROCEDURE

Starting from the expansion (2.20) we now describe the policy improvement procedure announced at the beginning of this chapter.

Let f be a decision function such that P(f) is quasi-compact, and let $g \in \mathbb{D}$. Then for every $n \in \mathbb{N}_{-1}$ we define real functions $\psi^n(f)$ and $\psi^n(g,f)$ on K and E, respectively, as follows:

$$\psi^{-1}(f)(x,d) := \sum_{y\in E}(p(x,d,y) - \delta_{xy})v_{-1}^f(y),$$

$$\psi^{0}(f)(x,d) := r(x,d) + \sum_{y\in E}(p(x,d,y) - \delta_{xy})v_{0}^f(y) - v_{-1}^f(x),$$

$$\psi^{n}(f)(x,d) := \sum_{y\in E}(p(x,d,y) - \delta_{xy})v_{n}^f(y) - v_{n-1}^f(x),$$

($n \in \mathbb{N}$, $(x,d) \in K$). Furthermore, let

$$\psi^n(g,f)(x) := \psi^n(f)(x,g(x))$$

for $x \in E$, $n \in \mathbb{N}_{-1}$. Hence, by reformulation, we have

$$\phi^{-1}(g,f) = \qquad (P(g) - I)v_{-1}^{f},$$
$$\phi^{o}(g,f) = r(g) + (P(g) - I)v_{o}^{f} - v_{-1}^{f},$$
$$\phi^{n}(g,f) = \qquad (P(g) - I)v_{n}^{f} - v_{n-1}^{f}$$

($n \in \mathbb{N}$). These function can be taken as a whole to form vector-valued functions $\Psi^n(f)$ and $\Psi^n(g,f)$, respectively, on K and E, respectively, for every $n \in \mathbb{N}_{-1}$. For $n \in N_{-1}$, $(x,d) \in K$ let

$$\Psi^n(f)(x,d) := (\phi^{-1}(f),\ldots, \phi^n(f))(x,d) := (\phi^{-1}(f)(x,d),\ldots, \phi^n(f)(x,d)),$$

and for $x \in E$

$$\Psi^n(g,f)(x) := (\phi^{-1}(g,f),\ldots, \phi^n(g,f))(x) := (\phi^{-1}(g,f)(x),\ldots, \phi^n(g,f)(x)).$$

Consequently, we can also write:

$$\Psi^n(f) = (\phi^{-1}(f),\ldots, \phi^n(f)) \text{ and } \Psi^n(g,f) = (\phi^{-1}(g,f),\ldots, \phi^n(g,f)),$$

and we set $\phi^{-3}(g,f) := \phi^{-2}(g,f) := 0$.

Note: Due to remark 2.15 it follows that the functions $\phi^{-1}(f)$, $\phi^{o}(f)$, and $\Psi^{o}(f)$ as well as $\phi^{-1}(g,f)$, $\phi^{o}(g,f)$, and $\Psi^{o}(g,f)$ coincide with the identically denoted functions of 2.4.

Let us recall the notations defined in 2.1:
For each policy $\pi = (f_o, f_1,\ldots) \in \mathfrak{R}_M$ it is

$$Q_o(\pi) := I \text{ and } Q_n(\pi) := Q_{n-1}(\pi)P(f_{n-1}),$$

$n \in \mathbb{N}$. If one applies the policy π, then the representation

$$v_\beta^\pi = \sum_{n=o}^{\infty} \beta^n Q_n(\pi) r(f_n) \qquad (2.23)$$

of the expected discounted total reward follows directly from the definition of v_β^π. Furthermore, let (g, π) denote the following composed policy $(g, \pi) := (g, f_o, f_1,\ldots) \in \mathfrak{R}_M$ for an arbitrary decision function.

Lemma 2.30. Let $g \in \mathbb{D}$ and $\pi = (f_o, f_1,\ldots) \in \mathfrak{R}_M$. Then

$$v_\beta^g - v_\beta^\pi = (I - \beta P(g))^{-1}(v_\beta^{(g,\pi)} - v_\beta^\pi) \qquad (2.24)$$

for $\beta \in [0,1)$.

Proof: By (2.4) and (2.23) we obtain

$$(I - \beta P(g))(v_\beta^g - v_\beta^\pi) = r(g) - (I - \beta P(g))v_\beta^\pi$$
$$= r(g) + \beta P(g) v_\beta^\pi - v_\beta^\pi$$
$$= v_\beta^{(g,\pi)} - v_\beta^\pi .$$

Since $I - \beta P(g)$ has a bounded inverse in $\mathfrak{L}(B(E))$, the above assertion is proved, because of $\|\beta P(g)\| = \beta < 1$. ∎

Lemma 2.31. Let $f,g \in \mathbb{D}$, and let $P(f)$ be quasi-compact. Then for each $\rho \in (0,\|H(f)\|^{-1})$ the equation

$$V_\beta^{(g,f^\infty)} - V_\beta^f = \sum_{n=-1}^{\infty} \rho^n \psi^n(g,f) \tag{2.25}$$

holds, where this series absolutely converges with respect to the norm topology on $B(E)$.

Proof: From theorem 2.27 it follows that

$$V_\beta^f = (1+\rho)\sum_{n=-1}^{\infty} \rho^n v_n^f,$$

hence

$$\begin{aligned} V_\beta^{(g,f^\infty)} - V_\beta^f &= r(g) + (\beta P(g) - I)V_\beta^f \\ &= r(g) + (P(g) - (1+\rho)I)\sum_{n=-1}^{\infty} \rho^n v_n^f \\ &= r(g) + \sum_{n=-1}^{\infty} \rho^n((P(g) - I)v_n^f - \rho\, v_n^f) \\ &= r(g) + \sum_{n=-1}^{\infty} \rho^n((P(g) - I)v_n^f - v_{n-1}^f) \\ &= \rho^{-1}\psi^{-1}(g,f) + \psi^o(g,f) + \sum_{n=o}^{\infty} \rho^n \psi^n(g,f). \end{aligned}$$

The absolute convergence of this series in $B(E)$ is a direct consequence of the absolute convergence of the series (2.20). ∎

The two preceding lemmas yield

Lemma 2.32. Let $f,g \in \mathbb{D}$ and let $P(f)$ and $P(g)$ be quasi-compact. Then, for every $n \in \mathbb{N}_{-2}$,

$$v_n^g - v_n^f = P(g)^*\psi^{n+1}(g,f) + \sum_{k=o}^{n+1}(-1)^k H(g)^{k+1}\psi^{n-k}(g,f). \tag{2.26}$$

Proof: From the identity $(I - \beta P(g))^{-1} = -(1+\rho)R(\rho,P(g) - I)$, which is valid for $\beta \in [0,1)$ and which can be verified as in the proof of theorem 2.27, one can conclude, using theorem 2.20, that

$$(I - \beta P(g))^{-1} = (1+\rho)(\rho^{-1}P(g)^* + \sum_{m=o}^{\infty}(-1)^m \rho^m H(g)^{m+1}).$$

This series converges absolutely in $\mathfrak{L}(B(E))$. For $\rho(|\rho| < \min(\|H(f)\|^{-1}, \|H(g)\|^{-1})$, due to theorem 2.27, the relation follows:

$$\Delta(\rho) := (1+\rho)\sum_{n=-1}^{\infty} \rho^n(v_n^g - v_n^f) = V_\beta^g - V_\beta^f.$$

From (2.24) and (2.25) it now results that

$$\begin{aligned} \Delta(\rho) &= (I - \beta P(g))^{-1}(V_\beta^{(g,f^\infty)} - V_\beta^f) \\ &= (1+\rho)(\rho^{-1}P(g)^* + \sum_{m=o}^{\infty}(-1)^m \rho^m H(g)^{m+1})(V_\beta^{(g,f^\infty)} - V_\beta^f) \end{aligned}$$

$$= (1+\rho)(\rho^{-1}P(g)^* + \sum_{m=0}^{\infty} (-1)^m \rho^m H(g)^{m+1})(\sum_{n=-1}^{\infty} \rho^n \psi^n(g,f)).$$

Since the two series (in parentheses) converge absolutely in $\mathfrak{L}(B(E))$ and in $B(E)$, respectively, we obtain that

$$\Delta(\rho) = (1+\rho)(\sum_{n=-1}^{\infty} \rho^{n-1} P(g)^*\psi^n(g,f) + \sum_{n=-1}^{\infty} \rho^n \sum_{k=0}^{n+1} (-1)^k H(g)^{k+1} \psi^{n-k}(g,f)),$$

and since $\psi^{-2}(g,f) = 0$

$$\Delta(\rho) = (1+\rho)\sum_{n=-1}^{\infty} \rho^n(P(g)^*\psi^{n+1}(g,f) + \sum_{k=0}^{n+1}(-1)^k H(g)^{k+1}\psi^{n-k}(g,f)).$$

If we compare the coefficients of the two representations of $\Delta(\rho)$, the assertion is proved for $n \in \mathbb{N}_{-1}$. It results directly from the definition of $\psi^{-1}(g,f)$ since $P(g)^*P(g) = P(g)^*$ if $n = -2$. ∎

Finally we make use of

<u>Lemma 2.33.</u> Let $f \in \mathbb{D}$, and let $P(f)$ be quasi-compact. Then $\Psi^N(f,f) = 0$ for all $N \in \mathbb{N}_{-1}$.

<u>Proof:</u> Due to theorem 2.27, theorem 1.3, and theorem 1.17 the equation

$$(P(f) - I)v_{-1}^f = (P(f) - I)P(f)^*r(f) = 0$$

holds and

$$(P(f) - I)v_n^f = (-1)^n(P(f) - I)\, H(f)^{n+1}\, r(f)$$
$$= (-1)^n(P(f)^* - I)\, H(f)^n\, r(f)$$
$$= \begin{cases} (-1)^{n-1}H(f)^n\, r(f) = v_{n-1}^f & \text{for } n \geq 1 \\ v_{-1}^f - r(f) & \text{for } n = 0. \end{cases}$$

Using these identities we obtain that $\psi^n(f,f) = 0$ for each $n \in \mathbb{N}_{-1}$.

Before stating the next theorem we think it useful to introduce some formalized notations.

Let $f, g \in \mathbb{D}$, and let $P(f)$ and $P(g)$ be quasi-compact. We denote the set of states which are recurrent under g by R. Furthermore, let $T := E \setminus R$. As in the proof of theorem 2.17, we define

$$R(x) := \{y \in R:\ x \overset{g}{\rightsquigarrow} y\},$$
$$T(x) := \{y \in T:\ x \overset{g}{\rightsquigarrow} y\}, \text{ and}$$
$$E(x) := R(x) \cup T(x).$$

Let $N \in \mathbb{N}_{-1}$ be an arbitrary but fixed number. For every $x \in E$ and $k = -1, \ldots, N+1$ we consider the following statements:

$C_k(x): \iff \psi^{-2}_{E(x)} = \dots = \psi^{k-1}_{E(x)} = 0 \wedge \psi^{k}_{R(x)} = 0,$

$A_k(x): \iff C_k(x) \wedge$ " $\psi^{k+1}_{R(x)} > 0$",

$B_k(x): \iff C_k(x) \wedge$ " $\psi^{k}_{T(x)} > 0$" [1]), and

$A^*(x): \iff \bigvee_{k=-1}^{N} A_k(x), \quad B^*(x): \iff \bigvee_{k=-1}^{N} B_k(x).$

Theorem 2.34. Let f and g be decision functions with P(f) and P(g) being quasi-compact. Moreover, let $\Psi^{N+1}(g,f) \succcurlyeq 0$. Then

(i) $\mathfrak{v}_N(g) \succcurlyeq \mathfrak{v}_N(f)$.

In detail the following equivalence relations are true for each $x \in E$:

(ii) $C_k(x) \iff \mathfrak{v}_{k-1}(g)(x) = \mathfrak{v}_{k-1}(f)(x) \wedge \mathfrak{v}_k(g)(x) \succcurlyeq \mathfrak{v}_k(f)(x)$ $(k = -1,\dots,N)$ and

(iii) $A_k(x) \vee B_k(x) \iff \mathfrak{v}_{k-1}(g)(x) = \mathfrak{v}_{k-1}(f)(x) \wedge \mathfrak{v}_k(g)(x) \not\succcurlyeq \mathfrak{v}_k(f)(x)$ $(k = -1,\dots,N)$. Furthermore, it holds that

(iv) $C_{N+1}(x) \iff \mathfrak{v}_N(g)(x) = \mathfrak{v}_N(f)(x)$, and

(v) $A^*(x) \vee B^*(x) \iff \mathfrak{v}_N(g)(x) \succ \mathfrak{v}_n(f)(x)$.

If the decision function g satisfies the additional assumption

$$\Psi^{N+1}(g,f)(x) = 0 \implies g(x) = f(x) \qquad (2.27)$$

for each $x \in E$, then, in addition, the following equivalence relations are valid:

(vi) $C_{N+1}(x) \iff \mathfrak{v}_N(g)(x) = \mathfrak{v}_N(f)(x) \wedge \mathfrak{v}_{N+1}(g)(x) \not\succ \mathfrak{v}_{N+1}(f)(x)$

and

(vii) $B_{N+1}(x) \iff \mathfrak{v}_N(g)(x) = \mathfrak{v}_N(f)(x), \mathfrak{v}_{N+1}(g)(x) \succ \mathfrak{v}_{N+1}(f)(x)$, and x is transient under g.

Proof: (For brevity, we shall write $\psi^{-1},\dots,\psi^{N+1}$, Ψ^{N+1} instead of $\psi^{-1}(g,f)$ etc., P^* instead of $P(g)^*$, and H instead of H(g).) Let $R_1,\dots,R_r$ be the classes of states which are recurrent under g, i.e., $R = \bigcup_{n=1}^{r} R_n$ (since P(g) is quasi-compact it holds that $1 \leq r < \infty$. Further, let $\Delta^n := v^g_n - v^f_n$, $n \in \mathbb{N}_{-1}$, and $\Delta^{-2} := 0$.

Firstly, we shall prove two auxiliary assertions:

1°: Let x be a recurrent state under g. (Without loss of generality, let $x \in R_1$.) Then, for all k = -1, ..., N+1, the equivalence relation

(a) $\Delta^{-2}(x) = \dots = \Delta^{k-1}(x) = 0 \iff \psi^{-1}_{R_1} = \dots = \psi^{k}_{R_1} = 0$

holds. Under the assumption $\Delta^{-2}(x) = \dots = \Delta^{k-1}(x) = 0$ it holds

(b) $\Delta^k(x) \geq 0$, and

[1]) In the case $T(x) = \emptyset$, we define " $\psi^{k}_{T(x)} > 0$": $\iff \emptyset$ (empty statement).

(c) $\Delta^k(x) > 0 \iff \psi_{R_1}^{k+1} > 0$

for all $k = -1, \ldots, N$. Under the additional assumption (2.27) it furthermore holds that

(d) $\psi_{R_1}^{-1} = \ldots = \psi_{R_1}^{N+1} = 0 \Rightarrow \Delta^{-1}(x) = \ldots = \Delta^{N+1}(x) = 0.$

Proof of 1°: (a) will be shown by induction with respect to n. For this purpose we remark that formula (2.26) takes the form

$$\Delta_{R_1}^n = (P_{R_1R_1})^* \psi_{R_1}^{n+1} + \sum_{m=0}^{n+1} (-1)^m (H_{R_1R_1})^{m+1} \psi_{R_1}^{n-m} \qquad (1^*)$$

in view of the lemma 1.5 and theorem 1.17. If $k = -1$, then (a) is almost trivial since $\psi_{R_1}^{-2} = 0$ (from this it follows " $\Leftarrow$ "), and due to (1*) it follows $P_{R_1R_1}^* \psi_{R_1}^{-1} = 0$. Remarking that $(P_{R_1R_1})^*$ only consists of positive elements and $\psi_{R_1}^{-1} = 0$ (due to the assumption) the identity $\psi_{R_1}^{-1} = 0$ must be valid (thus " $\Rightarrow$ " is shown).
Now let (a) hold for any $k \in \mathbb{N}$. If $\Delta^{-2}(x) = \ldots = \Delta^{k-1}(x) = \Delta^k(x) = 0$, then $\psi_{R_1}^{-1} = \ldots = \psi_{R_1}^{k} = 0$ by assumption. Since $\Psi^{N+1} \succcurlyeq 0$ it follows that $\psi_{R_1}^{k+1} \geqq 0$. (1*) implies $((P_{R_1R_1})^* \psi_{R_1}^{k+1})(x) = 0$. This is only possible if $\psi_{R_1}^{k+1}$ vanishes. In contrast, if $\psi_{R_1}^{-1} = \ldots = \psi_{R_1}^{k+1} = 0$ this can be inserted into (1*) implying $\Delta^{-2}(x) = \ldots = \Delta^k(x) = 0$ by the assumption. This proves (a).
Now let k be arbitrary with $-1 \leqq k \leqq N$, and let $\Delta^{-2}(x) = \ldots = \Delta^{k-1}(x) = 0$. From (a) and (1*) it follows $\psi_{R_1}^{k+1} \geqq 0$ and $\Delta^k(x) = ((P_{R_1R_1})^*)(x) \geqq 0$, where the following equivalence relation obviously holds: $\Delta^k(x) > 0 \iff \psi_{R_1}^{k+1} > 0$. This proves (a) and (b).
About (d):
From $\psi_{R_1}^{-1} = \ldots = \psi_{R_1}^{N+1} = 0$ and (2.27) it results that $g_{R_1} = f_{R_1}$ implying $\psi^{N+1}(g,f)_{R_1} = \psi^{N+1}(f,f)_{R_1} = 0$ in accordance with lemma 2.33. (1*) then provides that $\Delta^{N+1}(x) = ((P_{R_1R_1})^* \psi_{R_1}^{N+2}(g,f))(x) = 0.$ ⌋

The next auxiliary assertion is

2°.: Let x be a transient state under g. Then, for every $k = -1, \ldots, N+1$,

(a) $\Delta^{-2}(x) = \ldots = \Delta^{k-1}(x) = 0 \iff \psi_{E(x)}^{-2} = \ldots = \psi_{E(x)}^{k-1} = 0 \wedge \psi_{R(x)}^{k} = 0;$

if, for all $k = -1, \ldots, N+1$,

$$\Delta^{-2}(x) = \ldots = \Delta^{k-1}(x) = 0,$$

then

(b) $\Delta^k(x) \geqq 0$ and

(c) $\Delta^k(x) > 0 \iff \psi_{R(x)}^{k+1} > 0 \vee \psi_{T(x)}^{k} > 0.$

Moreover, under the additional assumption (2.27), it holds that

(d) $\psi^{-1}_{E(x)} = \dots = \psi^{N}_{E(x)} = 0 \wedge \psi^{N+1}_{R(x)} = 0$

$$\Longrightarrow \Delta^{-1}(x) = \dots = \Delta^{N}(x) = 0 \wedge \Delta^{N+1}(x) \geq 0,$$

as well as

(e) $\Delta^{N+1}(x) > 0 \Longleftrightarrow \psi^{N+1}_{T(x)} > 0$ provided that

$\psi^{-1}_{E(x)} = \dots = \psi^{N}_{E(x)} = 0 \wedge \psi^{N+1}_{R(x)} = 0.$

Proof of 2°: (2.26) implies

$$\Delta^{n}_{T}(x) = ((P_{TR(x)})^{*}\psi^{n+1}_{R(x)} + \sum_{m=0}^{n+1}(-1)^{m}(H_{TR(x)})^{m+1}\psi^{n-m}_{R(x)}$$

$$+ \sum_{m=0}^{m+1}(-1)^{m}(H_{TT(x)})^{m+1}\psi^{n-m}_{R(x)})(x), \qquad (2^{*})$$

$n \in \mathbb{N}_{-1}$, by using theorem 1.17.

Now we prove (a) by induction with respect to k. For $k = -1$ there is nothing to prove because the corresponding part of the proof for 1° shows that $\Delta^{-2}(x) = 0$ and $\psi^{-1}_{R(x)} = 0$. Now suppose (a) to be true for an arbitrary $k \leq N$. If $\Delta^{-2}(x) = \dots = \Delta^{k-1}(x) = \Delta^{k}(x) = 0$, it follows from the assumption that $\psi^{-2}_{E(x)} = \dots = \psi^{k-1}_{E(x)} = 0$ and $\psi^{k}_{R(x)} = 0$ and, from $\Psi^{N+1} \geqslant 0$, $\psi^{k+1}_{R(x)} \geq 0$ and $\psi^{k}_{T(x)} \geq 0$. (2*) then provides

$$\Delta^{k}_{T(x)} = ((P_{TR(x)})^{*}\psi^{k+1}_{R(x)} + H_{TT(x)}\psi^{k}_{T(x)})(x). \qquad (3^{*})$$

Since all elements of the x-row of $(P_{TR(x)})^{*}$ and $H_{TT(x)}$ are positive (cf. theorems 1.3 and 1.17) and $\Delta^{k}_{T(x)} = 0$ it must hold that $\psi^{k+1}_{R(x)} = 0$ and $\psi^{k}_{T(x)} = 0$.

If, on the contrary, $\psi^{-1}_{E(x)} = \dots = \psi^{k}_{E(x)} = 0$ and $\psi^{k+1}_{R(x)} = 0$, then, from the assumption and due to (2*) one obtains $\Delta^{-2}(x) = \dots = \Delta^{k}(x) = 0$. Thus (a) is proved.

Let k be arbitrary with $-1 \leq k \leq N$ and $\Delta^{-2}(x) = \dots = \Delta^{k-1}(x) = 0$. Then (3*) holds as has just been shown. From (a) and $\Psi^{N+1} \geqslant 0$ it follows $\psi^{k}_{T(x)} \geq 0$ and $\psi^{k+1}_{R(x)} \geq 0$. Hence, by (3*), holds $\Delta^{k}(x) \geq 0$ holds thus proving the equivalence relation

$$\Delta^{k}(x) > 0 \Longleftrightarrow \psi^{k}_{T(x)} > 0 \vee \psi^{k+1}_{R(x)} > 0.$$

Consequently, (b) and (c) are proved.

Concerning (d) and (e): If $\psi^{-1}_{E(x)} = \dots \psi^{N}_{E(x)} = 0$ and $\psi^{N+1}_{R(x)} = 0$, in view of $\Psi^{N+1}(g,f) \geqslant 0$ and (2.27), it follows $g_{R(x)} = f_{R(x)}$ (implying $\psi^{N+2}_{R(x)}(g,f) = \psi^{N+2}_{R(x)}(f,f) = 0$ due to lemma 2.33) and $\psi^{N+1}_{T(x)} \geq 0$. (3*) then yields $\Delta^{N+1}_{T(x)}(x) = H_{TT(x)}\psi^{N+1}_{T(x)}(x)$. Since the x-row of $H_{TT(x)}$ only consists of positive elements it follows that $\Delta^{N+1}(x) \geq 0$ (establishing (d)) and: $\Delta^{N+1}(x) > 0 \Leftrightarrow \psi^{N+1}_{T(x)} = 0$ (the latter shows (e)). ⌋

Now let $x \in E$ be arbitrarily chosen. Firstly, we assume x to be recurrent under g (for instance, $x \in R_1$). Then $R(x) = E(x) = R_1$; this implies that $B_k(x)$ is always wrong. Therefore, 1°(a) and (b) imply (ii),

(iii) follows from 1^o(a) and (c), and (vi) is a consequence of 1^o(a) and (d). Part (vii) is trivial since both sides are false.
On the other hand, if x is transient, (ii), (iii), (iv), and (vi) follow from 2° in a completely analogous manner. (vii) follows from 2^o(a) and (e).

It remains to show that (i) and (v) are valid. Obviously, it holds that

$$C_{N+1}(x) \Leftrightarrow \neg\ (A^*(x) \vee B^*(x))$$

for each $x \in E$. Hence, for every $x \in E$, it holds either $C_{N+1}(x)$ (and, consequently, $\mathfrak{v}_N(g)(x) = \mathfrak{v}_N(f)(x)$) or, for a certain $k \leqq N$, $A_k(x) \vee B_k(x)$ (this implies $\mathfrak{v}_k(g)(x) \succ \mathfrak{v}_k(f)(x)$). Thus the proof is complete. ∎

Theorem 2.34 plays an analogous role with respect to sensitive discount criteria as theorem 2.17 with regard to the average reward criterion. Again we shall prove some simple consequences.

Corollary 2.35. Let $f, g \in \mathbb{D}$, let $P(f)$ and $P(g)$ be quasi-compact, and let $\Psi^{N+1}(g,f) \succcurlyeq 0$ for a certain $N \in \mathbb{N}_{-1}$. Then the implication
(i) $\Psi^N(g,f)(x) \succ 0 \Rightarrow \mathfrak{v}_N(g)(x) \succ \mathfrak{v}_N(f)(x)$
is valid for every state $x \in E$, and for each state $x \in E$ which is recurrent under g it holds that
(ii) $\Psi^{N+1}(g,f)(x) \succ 0 \Rightarrow \mathfrak{v}_N(g)(x) \succ \mathfrak{v}_N(f)(x)$.

Proof: If, for any $x \in E$, $\Psi^N(g,f) \succ 0$, then the assertion $A_k(x) \vee B_k(x)$ must be true for some $k \leqq N$. If $\Psi^{N+1}(g,f)(x) \succ 0$ for a state x, which is recurrent under g, there is a $k \leqq N$ with the property that $A_k(x)$ is true. The statement then follows from theorem 2.34 (v) in both cases.

Let $f \in \mathbb{D}$. If $P(f)$ is quasi-compact, we define

$$\mathfrak{G}_N(f) := \{g \in \mathbb{D}:\ \Psi^N(g,f) \succ 0\} \qquad (N \in \mathbb{N}_{-1}).$$

(The set $\mathfrak{G}_o(f)$ coincides with the set $\mathfrak{G}_o(f)$ introduced in 2.4 as shown in remark 2.15.)

Corollary 2.36. Let $f, g \in \mathbb{D}$, let us assume $P(f)$ and $P(g)$ to be quasi-compact, and let $g \in \mathfrak{G}_{N+1}(f)$ for an $N \in \mathbb{N}_{-1}$. If

$$\Psi^{N+1}(g,f) = 0 \Rightarrow g(x) = f(x) \qquad (2.27)$$

for all $x \in E$, then it holds either

$$\mathfrak{v}_N(g) \succ \mathfrak{v}_N(f)$$

or

$$\mathfrak{v}_N(g) = \mathfrak{v}_N(f) \wedge \mathfrak{v}_{N+1}(g) \succ \mathfrak{v}_{N+1}(f).$$

Proof: Due to theorem 2.25 it holds that $\mathfrak{v}_N(g) \geqslant \mathfrak{v}_N(f)$. If $\mathfrak{v}_N(g) = \mathfrak{v}_N(f)$, then the assertion $C_{N+1}(x)$ is true for each $x \in E$ by theorem 2.34 (vi). Since, however, $\Psi^{N+1}(g,f) \neq 0$ there must be a state x_o such that $B_{N+1}(x_o)$ is true. Part (vii) of theorem 2.34 provides that $\mathfrak{v}_{N+1}(g)(x_o) > \mathfrak{v}_{N+1}(f)(x_o)$. Therefore, part (vi) of the same theorem implies $\mathfrak{v}_{N+1}(g) \succ \mathfrak{v}_{N+1}(f)$. ∎

Notes

1. In the case of a finite state space (in which P(f) and P(g) are, of course, quasi-compact) and N = -1 and N = 0, respectively, this statement is due to VEINOTT [117].

2. By significantly using the finiteness of E, MILLER and VEINOTT [84] proved the following result:
Let $N :=$ card E and $g \in \mathfrak{G}_N(f)$. Then it holds that $\mathfrak{v}_N(g) \succ \mathfrak{v}_N(f)$.

3. Using a more incisive condition the assertions of the corollaries 2.35 and 2.36 were obtained by KADOTA [66] for the special case N = -1.

4. The statements of theorem 2.34, corollary 2.35, and corollary 2.36, respectively, are identical with those of theorem 2.17, corollary 2.18, and corollary 2.19, respectively, in the case N = -1 if the assumptions used there are replaced by the quasi-compactness of P(f) and P(g). (In this context the statements C(x), A(x), and B(x), respectively, correspond to the statements $C_o(x)$, $A_{-1}(x)$, and $B_{-1}(x)$, respectively. As already said, the functions g^f and v^f, respectively, are equal to v^f_{-1} and v^f_o, respectively.)

Since both the assumptions used in 2.4 and those of the present section are automatically fulfilled if the state space is finite, theorem 2.34 is a generalization of theorem 2.17 in this case. The same can be said about the corollaries mentioned above.

Under more general assumptions theorem 2.17 and theorem 2.34 can be understood as generalizations of one and the same fact only differing in their directions.

Corollary 2.37. Let $f,g \in D$, let P(f) and P(g) be quasi-compact and let, for an $N \in \mathbb{N}_{-1}$, $\Psi^{N-1}(g,f) = 0$. Then it holds that [1])
(i) $\mathfrak{v}_{N-2}(g) = \mathfrak{v}_{N-2}(f)$ and
(ii) $g \in \mathfrak{G}_N(f) \cap \mathfrak{G}_{N+1}(f) \Rightarrow \mathfrak{v}_N(g) \succ \mathfrak{v}_N(f)$.

Proof: If $\Psi^{N-1}(g,f) = 0$ for each state $x \in E$, the assertion is true. Therefore, from theorem 2.34 (iv) (i) follows: $\mathfrak{v}_{N-2}(g) = \mathfrak{v}_{N-2}(f)$.
About (ii):

---

[1]) We set $\mathfrak{v}_{-3}(f) := \mathfrak{v}_{-2}(f) := 0$ and $\Psi^{-2}(g,f) := 0$ (null vector); $C_{-2}(x)$ denotes the empty statement.

The assumption $g \in \mathfrak{G}_N(f) \cap \mathfrak{G}_{N+1}(f)$ implies $\Psi^{N+1}(g,f) \succ 0$ (according to theorem 2.34 (i) it holds $\mathfrak{v}_N(g) \geqslant \mathfrak{v}_n(f)$). For a certain state $x_o \in E$ it holds $\Psi^N(g,f)(x_o) \succ 0$. Part (i) of corollary 2.35 provides that $\mathfrak{v}_N(g)(x_o) \succ \mathfrak{v}_n(f)(x_o)$. Together with the above relation (i) this proves the assertion $\mathfrak{v}_N(g) \succ \mathfrak{v}_N(f)$. ∎

Note: This statement is due to VEINOTT [116] for the finite state case. In [116] also the two cases were investigated in which either all states are recurrent or all states are transient, and assertions analogous to those of corollary 2.35 (iii) and theorem 2.34 (vii), respectively, were given. (In the second case the assumption (2.27) is not necessary since no recurrent state does exist. Such situations are not considered in the present paper.)

Furthermore, theorem 2.34 implies the following

Corollary 2.38. Let (V*) be fulfilled, and let a decision function $f^* \in \mathbb{D}$ be N-optimal on $\mathfrak{R}_S$ ($N \in \mathbb{N}_{-1}$). Then every decision function $g \in \mathbb{D}$ satisfying $\Psi^{N+1}(g,f^*) = 0$ is also N-optimal on $\mathfrak{R}_S$.

Proof: If, for a $g \in \mathbb{D}$, $\Psi^{N+1}(g,f^*) = 0$, then it holds $\mathfrak{v}_N(g) \geqslant \mathfrak{v}_N(f)$ according to part (i) of theorem 2.34. From lemma 2.28 it follows $\mathfrak{v}_N(g) \geqslant \mathfrak{v}_N(f^*) \geqslant \mathfrak{v}_N(h)$ for each $h \in \mathbb{D}$ since f^* is N-optimal on $\mathfrak{R}_S$. Thus g is also N-optimal on $\mathfrak{R}_S$. ∎

Theorem 2.34 justifies the policy improvement procedure announced above, which improves stationary policies step by step with regard to the criterion of N-optimality for a given $N \in \mathbb{N}_{-1}$. This procedure is complete analogously to the procedure of 2.4. In the following description of the policy improvement procedure we assume, for simplicity, that the assumption (V*) is fulfilled. Above all, for a given decision function $f \in \mathbb{D}$ the functions $\mathfrak{v}^f_{-1}, \dots, \mathfrak{v}^f_{N+1}$ have to be calculated. This can be done by using two methods, firstly, by using the formulas (2.21) and (2.22) (this requires $P(f)^*$ and $H(f)$ to be known), and secondly, by solving a system of linear equations of the following form.

Lemma 2.39. Let $N \in \mathbb{N}_{-1}$, $f \in \mathbb{D}$, and let $P(f)$ be quasi-compact. The functions $\mathfrak{v}^f_{-1}, \dots, \mathfrak{v}^f_{N+1}$ are the unique solutions of the system of the equations

$$(I - P(f))\mathfrak{v}_{-1} = 0, \tag{2.28}$$

$$(I - P(f))\mathfrak{v}_o = r(f) - \mathfrak{v}_{-1}, \tag{2.29}$$

$$(I - P(f))\mathfrak{v}_n = - \mathfrak{v}_{n-1}, \tag{2.30}$$

for $n = 1,\dots,N+1$, and

$$P(f)^*\mathfrak{v}_{N+1} = 0 \tag{2.31}$$

in $\mathfrak{v}_{-1}, \dots, \mathfrak{v}_{N+1}$.

Proof: Firstly, we show that $v_{-1}^f, \ldots, v_{N+1}^f$ solve (2.28) - (2.31). The validity of (2.28) and (2.29) was already shown in 2.4 (see remark 2.15). For $n \geq 1$, from (2.22) and theorem 1.17 we have

$$\begin{aligned}(I - P(f))v_n^f &= (-1)^n(I - P(f))H(f)^{n+1}\, r(f) \\ &= (-1)^n(I - P(f)^*)H(f)^n\, r(f) \\ &= -(-1)^n H(f)^n\, r(f) = -\, v_{n-1}^f\end{aligned}$$

as well as

$$P(f)^* v_n^f = (-1)^n P(f)^* H(f)^{n+1}\, r(f) = 0,$$

i.e., (2.30) and (2.31) are also fulfilled. In the following it can be seen that the solution is uniquely defined:

Let $v_{-1}, \ldots, v_{N+1}$ be bounded functions solving (2.28) - (2.31). Then, by lemma 2.9, it holds that $v_{-1} = v_{-1}^f$. It follows that

$$(I - P(f))(v_o - v_o^f) = 0$$

and

$$(I - P(f))(v_n - v_n^f) = -(v_{n-1} - v_{n-1}^f)$$

for $1 \leq n \leq N + 1$. By induction one obtains

$$(I - P(f))^{N+2}(v_{N+1} - v_{N+1}^f) = 0;$$

this yields

$$v_{N+1} - v_{N+1}^f \in \ker(I - P(f))^{N+2} = \ker(I - P(f)^*)$$

with regard to theorem 1.7. Finally, it follows from (2.31) that

$$v_{N+1} - v_{N+1}^f = P(f)^*(v_{N+1} - v_{N+1}^f) = 0.$$

Backward induction yields $v_n = v_n^f$ for $n = 0, \ldots, N+1$.

Note: If we want to avoid not only the calculation of $H(f)$ but also that of $P(f)^*$, then it suffices to solve (2.28), (2.29), and (2.30) for $n = 1, \ldots, N+2$. Analogous considerations show that

$$v_{-1} = v_{-1}^f, \ \ldots, \ v_{N+1} = v_{N+1}^f \, .$$

A similar method and efficient ways to solve the equations were given by VEINOTT [116] for the finite case. ∎

After having calculated $v_{-1}^f, \ldots, v_{N+1}^f$ a decision function $g \in \mathfrak{G}_{N+1}(f)$ must be chosen if $\mathfrak{G}_{N+1}(f) \neq \emptyset$. According to the theorem 2.34 (i) and lemma 2.28 g is not "worse" than f with respect to the criterion of N-optimality. Then the procedure can be continued with g instead of f. On the contrary, if $\mathfrak{G}_{N+1}(f) = \emptyset$, the iteration is stopped since it holds the following assertion:

Theorem 2.40.

(i) Let $f, g \in D$, and let $P(f)$ and $P(g)$ be quasi-compact. If, for any $N \in N_{-1}$, $\Psi^{N+1}(g,f) \;\; 0$, then
$\upsilon_N(g) \preccurlyeq \upsilon_n(f)$.

(ii) Let $N \in \mathbb{N}_{-1}$. If (V^*) is fulfilled and $\mathfrak{G}_{N+1}(f^*) = \emptyset$ for a decision function $f^* \in \mathbb{D}$, then $(f^*)^\infty$ is N-optimal on $\mathcal{R}_S$.

Proof: Part (i) can be proved completely analogous to part (i) of theorem 2.34.

About (ii):

Let $g \in D$ be arbitrarily chosen. Suppose the set

$$Mg := \{x \in E: \Psi^{N+1}(g,f^*)(x) \succ 0\}$$

to be non-empty. Then for the decision function $\bar{f}$ defined by

$$\bar{f}(x) := \begin{cases} f^*(x) & x \notin Mg \\ g(x) & x \in Mg \end{cases},$$

it would obviously hold that $\bar{f} \in \mathfrak{G}_{N+1}(f^*)$; this is in contradiction with the assumption. Hence $\Psi^{N+1}(g,f^*) \prec 0$ and, due to (i), $v_N(g) \preccurlyeq v_N(f^*)$. The application of lemma 2.28 completes the proof.

Notes:

1. If the assumption of (ii) is somewhat sharpened, it can be shown that $(f^*)^\infty$ is even N-optimal on $\mathcal{R}_M$. Such propositions will be given in the next section.

2. Although $\mathfrak{G}_{N+1}(f^*) = \emptyset$ is a stronger condition than N-optimality of $(f^*)^\infty$ on $\mathcal{R}_S$ (cf. remark 2.13), in general, we cannot derive from it that $(f^*)^\infty$ is (N+1)-optimal (on $\mathcal{R}_S$). This can easily be illustrated by examples.

3. It is not clear whether quasi-compactness of every P(f), $f \in \mathbb{D}$, implies the existence of a decision function $f^* \in \mathbb{D}$ satisfying $\mathfrak{G}_{N+1}(f^*) = \emptyset$.

The policy improvement procedure just introduced could be discussed in detail in the same way as it was done for the procedure of 2.4. Since both procedures are analogous we confine ourselves to the following remarks:

For obvious reasons we choose the new iterated $g \in \mathfrak{G}_{N+1}(f)$ following the point of view described in remark 2.12. By a repeated application of this version of the procedure a sequence of decision functions, called $(f_n)_{n=0,1,\ldots}$, is generated with $v_N(f_0) \precnsim v_N(f_1) \precnsim \ldots$ Since by corollary 2.36 cycles are excluded a direct consequence is the

Corollary 2.41. (VEINOTT [116]) Let $N \in \mathbb{N}_{-1}$. If the state space E is finite, then, for every initial function $f \in \mathbb{D}$, the policy improvement procedure stops after finitely many steps having found a decision function $f^* \in D$ which satisfies $\mathfrak{G}_{N+1}(f^*) = \emptyset$.

On the contrary, if the state space is countable the question again arises, whether the sequence $(f_n)_{n=0,1,\ldots}$ generated by the procedure (or a sequence of it) converges to an N-optimal decision function. As far as the authors know this question could not be completely answered yet.

Remark 2.16. We further mention here the following notion of optimality which is due to BLACKWELL [9]:
A policy $\pi^* \in \mathfrak{R}$ is called B-optimal if there is a $\beta_0 \in [0,1)$ with the property that, for each $\pi \in \mathfrak{R}$ and $\beta \in (\beta_0, 1)$, the inequality $V_\beta^{\pi^*} \geqq V_\beta^\pi$ holds. Obviously, a B-optimal policy is N-optimal for each $N \in \mathbb{N}_{-1}$.
If the finiteness of the state space E is supposed, then every policy being (card E)-optimal on $\mathfrak{R}_S$ is also N-optimal on $\mathfrak{R}_S$ for each $N \in \mathbb{N}_{-1}$ (MILLER and VEINOTT [84]). Hence, such a policy is B-optimal on $\mathfrak{R}_S$ as can be seen easily. Using theorem 2.4 we can prove that every such policy is B-optimal on $\mathfrak{R}$. Therefore, the propositions proved in connection with theorem 2.34 do not only show the existence of a stationary B-optimal policy (which was proved by BLACKWELL by using another approach) but also provide a finite procedure for the calculation of such a policy, whose applicability was first proved by MILLER and VEINOTT [84] and VEINOTT [116], respectively. The properties of the procedure described in these papers can be derived from theorem 2.34 and its corollaries (except for such essentially basing on the finiteness of E).

Note: If E is finite one can dispense with the expensive calculation of $v_{-1}^f, \dots, v_{N+1}^f$ $(N \geqq 1)$. Instead of this one can proceed as follows: Since, by lemma 2.28, every N-optimal stationary policy is also optimal for each n $(n \leqq N)$ first a (-1)-optimal decision function f_{-1}^* is calculated, then starting from f_{-1}^* a 0-optimal, and so on.

Only the two coefficients which are highest at the moment of calculation have to be computed for every step of the procedure.

2.6. SENSITIVE OPTIMALITY OF MARKOV POLICIES

K. SLADKY [106] derived a criterion for a Markov policy to be N-optimal on $\mathfrak{R}_M$. For this purpose he assumed that the state space is finite.

In this section we generalize the characterization given by SLADKY to the countable case. Moreover, the assertion of theorem 2.40(i) is strengthened, while at the same time more stringent assumptions are used. Thus an optimality statement on $\mathfrak{R}_M$ is derived. Finally, we quote some statements concerning the existence of sensitive optimal policies.

In this context we make use of the following condition:

(V**) For every $\varepsilon > 0$ there is a finite set $A_\varepsilon \subset E$ such that

$$P(f)(x, A_\varepsilon) \geqq 1 - \varepsilon$$

for every $x \in E$ and $f \in \mathbb{D}$.

Obviously, under (V**) every transition operator P(f), $f \in \mathbb{D}$ satisfies a DOEBLIN-condition. I.e., P(f) is quasi-compact. Simple examples show that the reversion is not true. Hence, (V^{**}) is properly sharper than (V^*). This is also expressed by the fact that all of the recurrence classes generated by the transition matrices $P(f)$, $f \in \mathbb{D}$, can be represented by a single system of representatives, which is finite. Finally, (V^{**}) implies a "uniform" recurrence in the following sense: For every $\varepsilon > 0$ and $k \in \mathbb{N}$ the k-th moment of the random first entrance time into the set A_ε is uniformly bounded in x and f, if the underlying process started from the state x and was ruled by the stationary policy f^∞, $f \in \mathbb{D}$. (This can be checked by simple calculations).

Notes:

1. (V^{**}) is a special case of the "simultaneous DOEBLIN condition": There are a number $\alpha > 0$, a finite set $A \subset E$, and a positive integer N satisfying

$$P(f)^N(x, A) \geqq \alpha$$

for every $x \in E$ and $f \in \mathbb{D}$. This condition was introduced by HORDIJK [55].

2. If E is finite, then (V^{**}) is trivially fulfilled.

Lemma 2.42. Let (V^{**}) be fulfilled. Then for every $\varepsilon > 0$ there is a finite set $A_\varepsilon \subset E$ such that

$$Q_n(\pi)(x, A_\varepsilon) \geqq 1 - \varepsilon$$

for every $n \in \mathbb{N}$, $x \in E$, and $\pi \in \mathfrak{R}_M$.

Proof: The statement is identical with (V^{**}) for n = 1. Now we suppose that the assertion is true for an arbitrary $n \in \mathbb{N}$. Let $\pi = (g_0, g_1, \ldots) \in \mathfrak{R}_M$ be arbitrary. Let $\pi' = (g_1, g_2, \ldots)$, then $Q_{n+1}(\pi) = P(g_0)Q_n(\pi')$, hence

$$Q_{n+1}(\pi)(x, A_\varepsilon) = \sum_{y \in E} P(g_0)(x,y)Q_n(\pi')(y, A_\varepsilon)$$

$$\geqq \sum_{y \in E} P(g_0)(x,y)(1 - \varepsilon) = 1 - \varepsilon,$$

($x \in E$). This proves the assertion for n + 1. ∎

The technique applied to the proofs of the following results essentially bases on the method developed by SLADKY for the finite case.

Lemma 2.43. Let $f \in \mathbb{D}$ and let $P(f)$ be quasi-compact. Further let $\pi = (g_0, g_1, \ldots) \in \mathfrak{R}_M$. For every $N \in \mathbb{N}$ and $\beta \in [0,1)$ let

$$V^\pi_{\beta,N} := \sum_{n=0}^{N-1} \beta^n Q_n(\pi)r(g_n).$$

Then, for each $\beta \in (0,1)$ and $N \geqq 2$,

$$V^{\pi}_{\beta,N} = (1-\beta)^{-1}[(1-\beta^N)v^f_{-1} + \sum_{n=o}^{N-2} (\beta^{n+1} - \beta^N)Q_n(\pi)\psi^{-1}(g_n,f)]$$
$$+ [\beta^{-1} v^f_o - \beta^N Q_N(\pi)v^f_o + \sum_{n=o}^{N-1} \beta^n Q_n(\pi)\psi^o(g_n,f)]$$
$$- (1 - \beta)\beta^{-1} \sum_{n=o}^{N} \beta^n Q_n(\pi)v^f_o.$$

Proof: Since

$$\psi^o(g,f) = r(g) + (P(g) - I)v^f_o - v^f_{-1},$$

($g \in \mathbb{D}$), (all vectors are well-defined because $P(f)$ is quasi-compact) we obtain

$$r(g) = \psi^o(g,f) + v^f_{-1} - (P(g) - I)v^f_o.$$

Together with the formula defining $V^{\pi}_{\beta,N}$, this equation imlies that

$$V^{\pi}_{\beta,N} = \sum_{n=o}^{N-1} \beta^n Q_n(\pi)\psi^o(g_n,f) + \sum_{n=o}^{N-1} \beta^n Q_n(\pi)v^f_{-1}$$
$$- \sum_{n=o}^{N-1} \beta^n Q_n(\pi)(P(g_n) - I)v^f_o \ . \qquad (1^*)$$

The second term of this sum can be rewritten in the following way: For $n \geqq 1$

$$Q_n(\pi) = Q_o(\pi) + \sum_{m=o}^{n-1} (Q_{m+1}(\pi) - Q_m(\pi)).$$

Let us consider the relation

$$Q_{m+1}(\pi) - Q_m(\pi) = Q_m(\pi)(P(g_m) - I).$$

Then the equations

$$(Q_{m+1}(\pi) - Q_m(\pi))v^f_{-1} = Q_m(\pi)(P(g_m) - I)v^f_{-1}$$
$$= Q_m(\pi)\psi^{-1}(g_m,f)$$

follow. Thus we obtain

$$\sum_{n=0}^{N-1} \beta^n Q_n(\pi)v^f_{-1}$$
$$= v^f_{-1} + \sum_{n=1}^{N-1} \beta^n(Q_o(\pi) + \sum_{m=o}^{N-1}(Q_{m+1}(\pi) - Q_m(\pi)))v^f_{-1}$$
$$= v^f_{-1} + \sum_{n=1}^{N-1} \beta^n(v^f_{-1} + \sum_{m=o}^{N-1} Q_m(\pi)\psi^{-1}(g_m,f)) =$$
$$= (\sum_{n=0}^{N-1} \beta^n)v^f_{-1} + \sum_{n=0}^{N-2} Q_n(\pi)\psi^{-1}(g_n,f)(\sum_{m=n+1}^{N-1} \beta^m)$$
$$= \frac{1 - \beta^N}{1 - \beta} v^f_{-1} + \sum_{n=0}^{N-2} \frac{\beta^{n+1} - \beta^N}{1 - \beta} Q_n(\pi)\psi^{-1}(g_n,f). \qquad (2^*)$$

Further, for the third term of (1*) we obtain

$$T_3 := \sum_{n=0}^{N-1} \beta^n Q_n(\pi)(P(g_n) - I)v_0^f = [\sum_{n=0}^{N-1} \beta^n Q_{n+1}(\pi) - \sum_{n=0}^{N-1} \beta^n Q_n(\pi)]v_0^f.$$

Now,

$$\sum_{n=0}^{N-1} \beta^n Q_n(\pi) = \beta \sum_{n=1}^{N-1} \beta^{n-1} Q_n(\pi) + Q_0(\pi) =$$

$$= \beta \sum_{n=1}^{N} \beta^{n-1} Q_n(\pi) - \beta^N Q_N(\pi) + Q_0(\pi) =$$

$$= \beta \sum_{n=0}^{N-1} \beta^n Q_{n+1}(\pi) - \beta^N Q_N(\pi) + Q_0(\pi),$$

hence

$$T_3 = [(1-\beta) \sum_{n=0}^{N-1} \beta^n Q_{n+1}(\pi) + \beta^N Q_N(\pi) - Q_0(\pi)]v_0^f$$

$$= [(1-\beta) \sum_{n=-1}^{N-1} \beta^n Q_{n+1}(\pi) + \beta^N Q_N(\pi) - \frac{1-\beta}{\beta} Q_0(\pi) - Q_0(\pi)]v_0^f$$

$$= \frac{1-\beta}{\beta} \sum_{n=0}^{N} \beta^n Q_n(\pi)v_0^f + \beta^N Q_N(\pi)v_0^f - \frac{1}{\beta} v_0^f. \qquad (3^*)$$

By inserting (2*) and (3*) into (1*) we obtain the asserted identity.∎

<u>Lemma 2.44.</u> Let $f \in \mathbb{D}$, let $P(f)$ be quasi-compact, and let $\pi = (g_0, g_1, \dots) \in \mathcal{R}_M$. Then, for all $m, N \in \mathbb{N}$,

$$\sum_{n=0}^{N+m-1} \beta^n Q_n(\pi)v_{m-1}^f = -\frac{1}{\beta} v_m^f + \beta^{N+m} Q_{N+m}(\pi)v_m^f - \sum_{n=0}^{N+m-1} \beta^n Q_n(\pi)\psi^m(g_n,f) + \frac{1-\beta}{\beta} \sum_{n=0}^{N+m} \beta^n Q_n(\pi)v_m^f$$

$(\beta \in (0,1))$.

<u>Proof:</u> For $m \in \mathbb{N}$ the equation

$$\psi^m(g,f) = (P(g) - I)v_m^f - v_{m-1}^f \quad (g \in \mathbb{D})$$

holds. This implies

$$\sum_{n=0}^{N+m-1} \beta^n Q_n(\pi)v_{m-1}^f = \sum_{n=0}^{N+m-1} \beta^n Q_n(\pi)(P(g_n) - I)v_m^f - \sum_{n=0}^{N+m-1} \beta^n Q_n(\pi)\psi^m(g_n,f). \qquad (1^*)$$

The first term of this sum satisfies the following relations:

$$\sum_{n=0}^{N+m-1} \beta^n Q_n(\pi)P(g_n) - I)v_m^f =$$

$$= \sum_{n=0}^{N+m-1} \beta^n Q_{n+1}(\pi)v_m^f - \sum_{n=0}^{N+m-1} \beta^n Q_n(\pi)v_m^f$$

$$= \sum_{n=0}^{N+m} \beta^{n-1} Q_n(\pi) v_m^f - \frac{1}{\beta} Q_o(\pi) v_m^f - \beta \sum_{n=0}^{N+m} \beta^{n-1} Q_n(\pi) v_m^f + \beta^{N+m} Q_{N+m}(\pi) v_m^f$$

$$= \frac{1-\beta}{\beta} \sum_{n=0}^{N+m} \beta^n Q_n(\pi) v_m^f - \frac{1}{\beta} v_m^f + \beta^{N+m} Q_{N+m}(\pi) v_m^f .$$

If this is inserted into (1*), we obtain the assertion. ∎

The two preceding lemmas lead to

Theorem 2.45. Let $f \in \mathbb{D}$, let $P(f)$ be quasi-compact and $\pi = (g_o, g_1, \ldots) \in \mathcal{R}_M$. Then, for all $\beta \in (0,1)$ and $M, N \in \mathbb{N}$ with $N \geq 2$,

$$v^{\pi}_{\beta,N} = \left(\frac{1-\beta}{\beta}\right)^{-1} \left[\frac{1-\beta^N}{\beta} v_{-1}^f + \sum_{n=0}^{N-2} (\beta^n - \beta^{N-1}) Q_n(\pi) \psi^{-1}(g_n, f)\right] +$$
$$+ \sum_{m=0}^{M-1} \left(\frac{1-\beta}{\beta}\right)^m \left[\frac{1}{\beta} v_m^f - \beta^{N+m} Q_{N+m}(\pi) v_m^f + \sum_{n=0}^{N+m-1} \beta^n Q_n(\pi) \psi^m(g_n, f)\right] -$$
$$- \left(\frac{1-\beta}{\beta}\right)^M \sum_{n=0}^{N+M-1} \beta^n Q_n(\pi) v_{M-1}^f . \tag{2.32}$$

Proof: Let N be fixed. Then (2.32) will be proved by induction with respect to M. If $M = 1$, then (2.32) is identical with

$$v^{\pi}_{\beta,N} = \frac{\beta}{1-\beta}\left[\frac{1-\beta^N}{\beta} v_{-1}^f + \sum_{n=0}^{N-2} (\beta^n - \beta^{N-1}) Q_n(\pi) \psi^{-1}(g_n, f)\right] +$$
$$+ \left[\frac{1}{\beta} v_o^f - \beta^N Q_N(\pi) v_o^f + \sum_{n=0}^{N-1} \beta^n Q_n(\pi) \psi^o(g_n, f)\right] -$$
$$- \frac{1-\beta}{\beta} \sum_{n=0}^{N-1} \beta^n Q_n(\pi) v_o^f .$$

The validity of this equation was proved in lemma 2.43. Now let us suppose that (2.32) is true for an arbitrary but fixed M. Then, due to lemma 2.44,

$$\sum_{n=0}^{N+m-1} \beta^n Q_n(\pi) v_{M-1}^f = - \frac{1}{\beta} v_M^f + \beta^{N+M} Q_{N+M}(\pi) v_M^f -$$
$$- \sum_{n=0}^{N+M-1} \beta^n Q_n(\pi) \psi^M(g_n, f) + \frac{1-\beta}{\beta} \sum_{n=0}^{N+M} \beta^n Q_n(\pi) v_m^f ,$$

and, consequently,

$$v^{\pi}_{\beta,N} = \left(\frac{1-\beta}{\beta}\right)^{-1} \left[\frac{1-\beta^N}{\beta} v_{-1}^f + \sum_{n=0}^{N-2} (\beta^n - \beta^{N-1}) Q_n(\pi) \psi^{-1}(g_n, f)\right] +$$
$$+ \sum_{m=0}^{M-1} \left(\frac{1-\beta}{\beta}\right)^m \left[\frac{1}{\beta} v_m^f - \beta^{N+m} Q_{N+m}(\pi) v_m^f + \sum_{n=0}^{N+m-1} \beta^n Q_n(\pi) \psi^m(g_n, f)\right] +$$
$$+ \left(\frac{1-\beta}{\beta}\right)^M \left[\frac{1}{\beta} v_M^f - \beta^{N+M} Q_{N+M}(\pi) v_M^f - \sum_{n=0}^{N+M-1} \beta^n Q_n(\pi) \psi^M(g_n, f)\right]$$
$$- \left(\frac{1-\beta}{\beta}\right)^{M+1} \sum_{n=0}^{N+M} \beta^n Q_n(\pi) v_M^f .$$

This, however, is just (2.32) for $M+1$. ∎

Corollary 2.46. Let $f \in \mathbb{D}$, let $P(f)$ be quasi-compact and $\pi = (g_0, g_1, \ldots) \in \mathcal{R}_M$. Then, for $\rho = \beta^{-1}(1-\beta)$, $\beta \in (0,1)$,

$$V_\beta^\pi = \sum_{m=-1}^{M-1} \rho^m(\beta^{-1} v_m^f + \sum_{n=0}^{\infty} \beta^n Q_n(\pi)\psi^m(g_n,f)) - \rho^M \sum_{n=0}^{\infty} \beta^n Q_n(\pi) v_{M-1}^f. \tag{2.33}$$

To prove this result we need

Lemma 2.47. Let $M \in \mathbb{N}$, $f \in \mathbb{D}$, and let $P(f)$ be quasi-compact. Then the family

$$\{\psi^m(g,f) \mid g \in \mathbb{D},\ m = -1,\ldots,M\}$$

is bounded in $B(E)$.

Proof: The quasi-compactness of $P(f)$ guarantees that all functions $\psi^m(g,f)$ exist and are elements of $B(E)$. Now, because $\|P(g)\| = 1$ and $\|r(g)\| < \tilde{k}$ for all $g \in \mathbb{D}$, one obtains that

$$\|\psi^{-1}(g,f)\| = \|(P(g) - I)v_{-1}^f\| \leqq 2\|v_{-1}^f\|$$

and

$$\|\psi^0(g,f)\| = \|r(g) + (P(g) - I)v_0^f - v_{-1}^f\| \leqq \tilde{k} + 2\|v_0^f\| + \|v_{-1}^f\|$$

as well as

$$\|\psi^m(g,f)\| \leqq 2\|v_m^f\| + \|v_{m-1}^f\|$$

for $m = 1,\ldots,M$. Hence,

$$\|\psi^m(g,f)\| \leqq \max\{\tilde{k} + 2\|v_m^f\| + 2\|v_{m-1}^f\| \mid m = -1,\ldots,M\}$$

for $m = -1,\ldots,M$ and $g \in \mathbb{D}$.

Proof of corollary 2.46: In view of (2.23) we obtain that $V_\beta^\pi = \lim_{N\to\infty} V_{\beta,N}^\pi$ in the sense of the norm convergence in $B(E)$.
Therefore, if we show that all the terms of (2.32) converge, then the assertion is proved. Both from this inequality

$$\|\beta^n Q_n(\pi)\psi^m(g_n,f)\| \leqq \beta^n \|\psi^m(g_n,f)\| \quad (n \in \mathbb{N},\ m = -1,\ldots,M-1)$$

and lemma 2.47 follows that

$$\lim_{N\to\infty} \sum_{n=0}^{N+m-1} \beta^n Q_n(\pi)\psi^m(g_n,f) = \sum_{n=0}^{\infty} \beta^n Q_n(\pi)\psi^m(g_n,f)$$

exists in $B(E)$. Furthermore, due to the inequalities

$$\|\sum_{n=0}^{N-2} \beta^{N-1} Q_n(\pi)\psi^{-1}(g_n,f)\| \leqq \beta^{N-1} \sum_{n=0}^{N-2} \|\psi^{-1}(g_n,f)\| \leqq$$

$$\leqq (N-1)\beta^{N-1} \sup_{g\in\mathbb{D}} \|\psi^{-1}(g,f)\| \leqq 2(N-1)\beta^{N-1}\|v_{-1}^f\|$$

and $\lim_{N\to\infty} (N-1)\beta^{N-1} = 0$, we obtain

$$\lim_{N\to\infty} \sum_{n=0}^{N-2} \beta^{N-2} Q_n(\pi)\psi^{-1}(g_n,f) = 0.$$

Finally, the inequality

$$\| \beta^n Q_n(\pi) v^f_{M-1} \| \leqq \beta^n \| v^f_{M-1} \|$$

implies the existence of

$$\lim_{N\to\infty} \sum_{n=0}^{N+M-1} \beta^n Q_n(\pi) v^f_{M-1} = \sum_{n=0}^{\infty} \beta^n Q_n(\pi) v^f_{M-1}.$$

Obviously, the remaining terms in (2.32) are also convergent if $N \to \infty$. Hence, (2.32) changes into (2.33) if $N \to \infty$. ∎

Note: If the state space E is finite, then we can dispense with the quasi-compactness assumption concerning P(f) in lemmas 2.43 and 2.44 as well as in theorem 2.45 and corollary 2.46, since it is automatically fulfilled. In this case the assertion just mentioned were proved by SLADKY [106].

Lemma 2.48. Let $m \in \mathbb{N}$, $f \in \mathbb{D}$, and let P(f) be quasi-compact. Further let

$$(\psi^m(g,f),\ \psi^{m+1}(g,f)) \preccurlyeq 0$$

for all $g \in \mathbb{D}$. Then for every finite set $A \subset E$ there is a constant $h_A > 0$ satisfying

$$-\ h_A\ \psi^{m+1}(g,f)_A \geqq \psi^m(g,f)_A$$

for every $g \in \mathbb{D}$.

Proof: Let $c := \max\{\psi^m(g,f)(x) < 0:\ g \in \mathbb{D},\ x \in A\}$ and $d := \max\{\psi^{m+1}(g,f)(x) > 0:\ g \in \mathbb{D},\ x \in A\}$ if these two sets are non-empty, otherwise let $c := -1$ and $d := 1$. (Note that the maxima are attained since A and the decision sets D(x), $x \in A$, are finite.) Then every constant $h_A \in (0, -\frac{c}{d}]$ satisfies the asserted inequality. If the two sets introduced above are non-empty this can be seen as follows: For each $x \in A$ either $\psi^m(g,f)(x) \leqq c < 0$ is valid (this implies $-h_A\ \psi^{m+1}(g,f)(x) \geqq c$ because $\psi^{m+1}(g,f)(x) \leqq d$ and $h_A \leqq -c^{-1}d$) or $\psi^m(g,f)(x) = 0$. In the second case, i.e., one of the above sets is empty $\psi^m(g,f)_A \leqq 0$ and $\psi^{m+1}(g,f)_A \leqq 0$. Hence, every constant $h_A \in (0,1]$ satisfies the relation

$$-\ h_A\ \psi^{m+1}(g,f)_A \geqq 0 \geqq \psi^m(g,f)_A\ . ∎$$

Now, we formulate the main result of this section.

Theorem 2.49. Let the assumption (V^{**}) be fulfilled. Let $N \in \mathbb{N}_{-1}$, $f^* \in \mathbb{D}$, and let us suppose that $\mathbb{O}_{N+1}(f^*) = \emptyset$. Then the following assertions hold:

(i) The stationary policy $(f^*)^\infty$ is N-optimal on $\mathfrak{K}_M$.

(ii) Any policy $\pi = (g_0, g_1, \dots) \in \mathfrak{K}_M$ is N-optimal on $\mathfrak{K}_M$, if and only if

$$Q_n(\pi)\psi^m(g_n, f^*) = 0 \tag{2.34}$$

and

$$\lim_{\beta \to 1} \frac{1-\beta}{\beta} \sum_{n=0}^{\infty} \beta^n Q_n(\pi)\psi^{N+1}(g_n, f^*) = 0 \tag{2.35}$$

for all $n \in \mathbb{N}_0$ and $m = -1, \dots, N$.

Proof:

I. Proof of (i):

The policy $(f^*)^\infty$ is N-discount optimal on $\mathfrak{K}_M$ if

$$\underline{\lim}_{\beta \to 1} \left(\frac{1-\beta}{\beta}\right)^{-N} (V_\beta^\pi - V_\beta^{f^*}) \leq 0$$

for all $\pi \in \mathfrak{K}_M$. We shall prove that this condition is fulfilled. For this purpose we consider the difference

$$\Delta_\rho := V_\beta^{f^*} - V_\beta^\pi .$$

The condition (V^{**}) implies that $P(f)$ is quasi-compact for every $f \in \mathbb{D}$. Therefore, the functions $v_n^{f^*} =: v_n^*$ and $\psi^m(g, f^*)$ $(m, n \in \mathbb{N}_{-1}, g \in \mathbb{D})$ are well-defined. For brevity, we write $Q_n := Q_n(\pi)$. Following theorem 2.27 we obtain

$$V_\beta^{f^*} = (1 + \rho)\left(\sum_{m=-1}^{N+1} \rho^m v_m^{f^*} + \sum_{m=N+2}^{\infty} \rho^m v_m^{f^*}\right)$$

and, in view of corollary 2.46,

$$V_\beta^\pi = \sum_{m=-1}^{N+1} \rho^m \left[\frac{1}{\beta} v_m^* + \sum_{n=0}^{\infty} \beta^n Q_n \psi^m(g_n, f^*)\right] - \rho^{N+2} \sum_{n=0}^{\infty} \beta^n Q_n v_{N+1}^*$$

for $0 < \rho < \|H(f^*)\|^{-1}$.

Since

$$(1 + \rho) \sum_{n=N+2}^{\infty} \rho^n v_n^* = o(\rho^{N+1}) \quad ^{1)}$$

for $\rho \to 0$ and

$$\left\|\rho^{N+2} \sum_{n=0}^{\infty} \beta^n Q_n v_{N+1}^*\right\| \leq \rho^{N+2} \frac{1}{1-\beta} \|v_{N+1}^*\| = \frac{\rho^{N+1}}{\beta} \|v_{N+1}^*\|$$

---

[1]) Here the Landau symbol $o(\rho^N)$ denotes a function $F: (0, \infty) \to B(E)$ with $\lim_{\rho \to 0} \rho^{-N} F(\rho) = 0$.

the following representation holds:

$$\Delta_\rho = - \sum_{m=-1}^{N+1} \rho^m \sum_{n=0}^{\infty} \beta^n Q_n \psi^m(g_n, f^*) + o(\rho^N). \qquad (1^*)$$

In the following we simply use:

$$\lambda_n^m := \psi^m(g_n, f^*).$$

For the first term in (1*) the assumption $\mathfrak{G}_{N+1}(f^*) = \emptyset$ implies that the "coefficient matrix"

$$M := \left(\sum_{n=0}^{\infty} \beta^n Q_n \psi^{-1}(g_n, f^*), \ldots, \sum_{n=0}^{\infty} \beta^n Q_n \psi^{N+1}(g_n, f^*)\right)$$

is lexicographically non-positive, i.e.,

$$M(x) = \left(\sum_{n=0}^{\infty} \beta^n Q_n \psi^{-1}(g_n, f^*)(x), \ldots, \sum_{n=0}^{\infty} \beta^n Q_n \psi^{N+1}(g_n, f^*)(x)\right) \preccurlyeq 0.$$

Applying the notation

$$E_n(x) := \{y \in E : Q_n(x,y) > 0\}$$

this can be seen as follows. Due to $\mathfrak{G}_{N+1}(f^*) = \emptyset$ we obtain

$$\psi^{-1}(g_n, f^*) = \lambda_n^{-1} \leqq 0$$

for all $n \in \mathbb{N}_0$. Hence

$$\sum_{n=0}^{\infty} \beta^n Q_n \lambda_n^{-1}(x) \leqq 0,$$

where "=" holds, if and only if $\lambda_n^{-1}|_{E_n(x)} = 0$ for all $n \in \mathbb{N}_0$. In the latter case, by $\mathfrak{G}_{N+1}(f^*) = \emptyset$, we again have that

$$\lambda_n^0|_{E_n(x)} \leqq 0$$

for $n \in \mathbb{N}_0$. This implies that

$$\sum_{n=0}^{\infty} \beta^n Q_n \lambda_n^0(x) \leqq 0.$$

In the "="-case we conclude that

$$\lambda_n^0|_{E_n(x)} \leqq 0$$

for $n \in \mathbb{N}_0$ in an analogous way. Continuing these conclusions we find $M(x) \preccurlyeq 0$ for all $x \in E$.

Henceforth, let $x \in E$ be arbitrary but fixed.
We define

$$m_0 := \begin{cases} \min\{m \in \{-1, \ldots, N+1\} : \exists\, n \in \mathbb{N}_0 : Q_n \lambda_n^m(x) < 0\} & \text{if } M(x) \prec 0 \\ N + 2 & \text{otherwise} \end{cases}$$

and, if $m_0 < N + 2$,

$$n_0 := \min\{n \in \mathbb{N}_0 : Q_n \lambda_n^{m_0}(x) < 0\}.$$

Then

$$\left.\begin{aligned}&\sum_{n=0}^{\infty} \beta^n Q_n \lambda_n^{m_0}(x) < 0\\ \text{and}\quad &\\ &\sum_{n=0}^{\infty} \beta^n Q_n \lambda_n^{m}(x) = 0 \quad (m < m_0)\end{aligned}\right\} \tag{2*}$$

for $m_0 \geqq 0$.
Moreover, if $n_0 > 0$, then due to the definition of n_0 we have

$$Q_n \lambda_n^{m_0}(x) = 0$$

for $n < n_0$. Therefore

$$Q_n \lambda_n^{m_0+1}(x) \leqq 0.$$

Considering (2*), the following formulas result from (1*):

$$\rho^{-m_0} \Delta_\rho(x) = \begin{cases} -\sum\limits_{m=m_0}^{m_0+1} \rho^{m-m_0} \sum\limits_{n=0}^{\infty} \beta^n Q_n \lambda_n^m(x) \\ -\sum\limits_{m=m_0+2}^{N+1} \rho^{m-m_0} \sum\limits_{n=0}^{\infty} \beta^n Q_n \lambda_n^m(x) + o(\rho^{N-m_0}) & (3^*) \\ \qquad\qquad \text{for } m_0 < N \\ -\sum\limits_{m=N}^{N+1} \rho^{m-N} \sum\limits_{n=0}^{\infty} \beta^n Q_n \lambda_n^m(x) + o(1) & (4^*) \\ \qquad\qquad \text{for } m_0 = N \end{cases}$$

$$\rho^{-N} \Delta_\rho(x) = \begin{cases} -\rho \sum\limits_{n=0}^{\infty} \beta^n Q_n \lambda_n^{N+1}(x) + o(1) \quad \text{for } m_0 = N+1 & (5^*) \\ O(1) \quad \text{for } m_0 > N+1. & (6^*) \end{cases}$$

In the following these four cases will be considered in detail.

1°. $\underline{m_0 < N}$: Since $\|Q_n(\pi)\| = 1$ for all $n \in \mathbb{N}_0$ and due to lemma 2.47 there is a constant $\tilde{H} \geqq 0$ with

$$\|Q_n \lambda_n^m\| \leqq \tilde{H} \quad \text{for } m = -1,\dots,N+1 \text{ and } n \in \mathbb{N}_0.$$

Therefore, we obtain, for $\rho \to 0$,

$$\left| \sum_{m=m_0+2}^{N+1} \rho^{m-m_0} \sum_{n=0}^{\infty} \beta^n Q_n \lambda_n^m(x) \right| \leqq (N-m_0)\rho^2 \tilde{H}(1-\beta)^{-1}.$$

The right hand term, however, is equal to $o(1)$ for $\rho \to 0$. Since the series

$$\sum_{n=0}^{\infty} \beta^n Q_n \lambda_n^m(x)$$

absolutely converges (3*) can be transformed into

$$\rho^{-m_0} \Delta_\rho(x) = -\sum_{n=0}^{\infty} \beta^n Q_n(\lambda_n^{m_0} + \rho\lambda_n^{m_0+1})(x) + o(1). \tag{7*}$$

For the first term in (7*)

$$T_\rho(x) := - \sum_{n=0}^{\infty} \beta^n Q_n(\lambda_n^{m_0} + \rho\lambda_n^{m_0+1})(x),$$

in view of the definitions of m_0 and n_0 one obtains

$$T_\rho(x) \geqq - \beta^{n_0} Q_{n_0}(\lambda_{n_0}^{m_0} + \rho\lambda_{n_0}^{m_0+1})(x)$$

$$- \sum_{m=n_0+1}^{\infty} \beta^n Q_n(\lambda_n^{m_0} + \rho\lambda_n^{m_0+1})(x). \qquad (8^*)$$

The limit of the second right hand term of this inequality

$$T'_\rho(x) := - \sum_{n=n_0+1}^{\infty} \beta^n Q_n(\lambda_n^{m_0} + \rho\lambda_n^{m_0+1})(x)$$

is nonnegative if ρ tends to zero as can be seen as follows:
Let $\varepsilon > 0$ be arbitrary. With regard to lemma 2.42 there is a finite set $A_\varepsilon \subset E$ with

$$Q_n(\pi)(x, A_\varepsilon) \geqq 1 - \varepsilon$$

for $n \in \mathbb{N}$. Due to lemma 2.48 we can find a constant $h_\varepsilon > 0$ with the property

$$\phi^{m_0}(g,f^*)_{A_\varepsilon} + \rho\phi^{m_0+1}(g,f^*)_{A_\varepsilon} \leqq 0$$

for all $g \in \mathbb{D}$ and $\rho \in (0,h_\varepsilon]$. In particular, we have that

$$(\lambda_n^{m_0} + \rho\lambda_n^{m_0+1})|_{A_\varepsilon} \leqq 0$$

for $\rho \in (0,h_\varepsilon]$ and $n \in \mathbb{N}$. Consequently,

$$T'_\rho(x) = - \sum_{n=n_0+1}^{\infty} \beta^n Q_n|_{A_\varepsilon}(\lambda_n^{m_0} + \rho\lambda_n^{m_0+1})|_{A_\varepsilon}(x)$$

$$- \sum_{n=n_0+1}^{\infty} \beta^n Q_n|_{\bar{A}_\varepsilon}(\lambda_n^{m_0} + \rho\lambda_n^{m_0+1})|_{\bar{A}_\varepsilon}(x)$$

$$\geqq - \sum_{n=n_0+1}^{\infty} \beta^n Q_n|_{\bar{A}_\varepsilon}(\lambda_n^{m_0} + \rho\lambda_n^{m_0+1})|_{\bar{A}_\varepsilon}(x).$$

Since $\lambda_n^{m_0} \leqq 0$ and $\|\lambda_n^{m_0+1}\| \leqq \tilde{H}$ for all n, we furthermore obtain that

$$T'_\rho(x) \geqq - \sum_{n=n_0+1}^{\infty} \beta^n Q_n|_{\bar{A}_\varepsilon} \rho \mathbf{1}_{\bar{A}_\varepsilon} \cdot \tilde{H} \geqq$$

$$- \rho \tilde{H} \varepsilon \frac{\beta^{n_0+1}}{1-\beta} = - \varepsilon \tilde{H} \beta^{n_0} \geqq - \varepsilon \tilde{H}. \qquad (9^*)$$

Consequently,

$$\varliminf_{\rho \downarrow 0} T'_\rho(x) \geqq 0.$$

Summarizing the results derived above and the fact that

$$Q_{n_0} \lambda_{n_0}^{m_0+1}(x) \leqq 0,$$

(8*) can be transformed into

$$\underline{\lim_{\rho\downarrow 0}}\, T_\rho(x) \geq - Q_{n_0}\lambda_{n_0}^{m_0}(x),$$

whereas we use the definitions of m_0 and n_0 and (9*). This relation implies in view of (7*) that

$$\underline{\lim_{\rho\downarrow 0}}\ \rho^{-m_0}\, \Delta_\rho(x) \geq - Q_{n_0}\, \lambda_{n_0}^{m_0}(x) > 0.$$

Since $m_0 < N$, finally, we obtain

$$\underline{\lim_{\rho\downarrow 0}}\, \rho^{-N}\, \Delta_\rho(x) > 0. \tag{10*}$$

This means that the policy $(f^*)^\infty$ is "better" than π in the state x, or otherwise: π is not N-optimal.

2^0. $\underline{m_0 = N}$: By arguments of absolute convergence (4*) can be immediately transformed into (7*). Analogous to the case 1^0 we derive (10*) from (7*).

3^0. $\underline{m_0 = N + 1}$: In view of the definition of m_0 the inequality

$$Q_n\, \lambda_n^{N+1}(x) \leq 0 \tag{11*}$$

holds for all $n \in N_0$. Hence, $\underline{\lim_{\rho\downarrow 0}}\ \rho^{-N}\, \Delta\rho(x) \geq 0$.

4^0. $\underline{m_0 > N + 1}$: Obviously, in this case the equation

$$\underline{\lim_{\rho\downarrow 0}}\, \rho^{-N}\Delta_\rho(x) = 0 \tag{12*}$$

is valid.

Summarizing the results of these four cases we obtain (11*) for every $x \in E$. This proves that the policy $(f^*)^\infty$ is N-optimal.

II. Proof of (ii):
A policy $\pi \in \mathfrak{R}_M$ is N-discount optimal on $\mathfrak{R}_M$, if and only if

$$\underline{\lim_{\rho\downarrow 0}}\, \rho^{-N}(V_\beta^\pi - V_\beta^{\pi'}) \geq 0$$

for every $\pi' \in \mathfrak{R}_M$. First we show that this is true if the conditions (2.34) and (2.35) are fulfilled. Using the notations of part I we consider the term

$$\rho^{-N}(V_\beta^\pi(x) - V_\beta^{f^*}(x)) = - \rho^{-N}\, \Delta_\rho(x).$$

If m_0 is defined as in part I, then $m_0 \geq N + 1$ follows immediately from (2.34). I.e., $\rho^{-N}\, \Delta_\rho(x)$ can be represented with respect to either (5*) or (6*). Note that the two terms are convergent if $\rho \to 0$. Due to (2.35) we directly obtain

$$\lim_{\rho\downarrow 0}(-\rho^{-N}\Delta_\rho(x)) = \lim_{\rho\downarrow 0}\rho^{-N}(V_\beta^\pi - V_\beta^{f^*}) = 0.$$

Therefore,

$$\underline{\lim}_{\rho\downarrow 0}\rho^{-N}(V_\beta^\pi - V_\beta^{\pi'}) \geq$$

$$\underline{\lim}_{\rho\downarrow 0}\rho^{-N}(V_\beta^\pi - V_\beta^{f^*}) + \underline{\lim}_{\rho\downarrow 0}\rho^{-N}(V_\beta^{f^*} - V_\beta^{\pi'}) \geq 0$$

for any policy $\pi' \in \mathfrak{R}_M$, i.e., π is N-discount optimal on $\mathfrak{R}_M$.

On the other hand, let $\pi \in \mathfrak{R}_M$ be N-discount optimal on $\mathfrak{R}_M$. In view of part I the equations

$$\underline{\lim}_{\rho\downarrow 0}\rho^{-N}(V_\beta^\pi - V_\beta^{f^*})(x) = \lim_{\rho\downarrow 0}(-\rho^{-N}\Delta_\rho(x)) = 0$$

must be valid for every $x \in E$. As has been shown in the first part this is only possible in the cases 3^0 and 4^0 where $m_0 \geq N + 1$. Due to the definition of m_0 (2.34) obviously holds. (2.35) can be derived from (5*) and (11*) in the case 3^0 for the chosen x. In the case 4^0 it is obvious. ∎

Notes:

1. Theorem 2.49 was proved by SLADKY [106] for finite state spaces. The basic ideas of the proof given here are similar to those of [106].

2. The assumption (V**) may seem to be very stringent. As already mentioned, it is trivially fulfilled in the finite case. With regard to the assertion (i) of theorem 2.49, similar statements were already given under weaker conditions but restricted to the case N = -1. HORDIJK and SLADKY [58] proved analogous statements for an arbitrary $N \in \mathbb{N}_{-1}$. Assumptions which exclude more than one class of recurrent states were required in all these cases. Thus other techniques of proving were applicable.
In view of part (ii) of theorem 2.49 concerning the characterization of N-optimal policies in $\mathfrak{R}_M$, the authors do not know any publications referring to the countable case.

3. By application to a stationary policy theorem 2.49 yields the following assertion: g^∞ is N-optimal on $\mathfrak{R}_M$, if and only if $\Psi^N(g,f^*) = 0$ and

$$\lim_{\rho\downarrow 0}\rho\sum_{n=0}^{\infty}\beta^n P(g)\,\psi^{N+1}(g,f^*) = 0.$$

In particular, g^∞ is N-optimal on $\mathfrak{R}_M$ if $\Psi^{N+1}(g,f^*) = 0$. This is a stronger statement than that of corollary 2.38; indeed, more incisive assumptions were applied in its proof.

NOTES CONCERNING THE EXISTENCE OF SENSITIVE-DISCOUNT OPTIMAL POLICIES

In connection with the generalized "HOWARD algorithm" of 2.5 as well as with respect to theorem 2.49, which is based on the existence of a stationary policy $(f^*)^\infty$ satisfying $\mathfrak{G}_{N+1}(f^*) = \emptyset$, those conditions, which guarantee that such a policy exists, are of interest. Corollary 2.41 shows that there is such a policy if the state space is finite. Moreover, remark 2.16 yields the following obvious consequence with regard to the fact that every N-optimal policy (on $\mathfrak{R}_S$) is also n-optimal (on $\mathfrak{R}_S$) for each $n \leq N$ (see corollary 2.29):

Theorem 2.50. Let $N \in \mathbb{N}_{-1}$. If the state space E is finite, then the stationary policy $(f^*)^\infty$ is N-optimal on $\mathfrak{R}$ for every decision function f^* satisfying $\mathfrak{G}_{N+1}(f^*) = \emptyset$.

In particular, theorem 2.49 can be used as an optimality criterion on $\mathfrak{R}$. The question whether a sensitive discount optimal policy does always exist under the supposition of (V*) or (V**) has not been answered yet in a complete generality if the state space is countable. A partial result for N = 0 was obtained by KADOTA [66], who made use of the following assumption:

(VUD): Let (V*) be fulfilled and suppose that there is a $B \in \mathbb{R}$ such that

$$\left\| \sum_{k=0}^{n} \left(P(f)^k - P(f^*)\right) \right\| \leq B \tag{2.36}$$

for every $f \in \mathbb{D}$ and $n \in \mathbb{N}_0$.

Theorem 2.51. (KADOTA [66]) Under the assumption (VUD) there is a stationary policy $(f^*)^\infty$ which is 0-discount-optimal and satisfies $\mathfrak{G}_0(f^*) = \emptyset$.

Note: There are the following relationships between the assumptions (V*), (V**) and (VUD):
The relation (2.36) coincides with the inequality (D*) of 1.2 for every fixed $f \in \mathbb{D}$ and it is a consequence of the quasi-compactness of P(f). (VUD) is more incisive than (V*), as far as the constant B in (2.36) must be independent of the choice of $f \in \mathbb{D}$. Consequently, it can be regarded as a "uniform DOEBLIN condition". [66] shows, however, that (VUD) does not imply (V**). The authors conjecture that, conversely, (VUD) does not follow from (V**), i.e., the conditions (V**) and (VUD) must be considered as non-equivalent strengthenings of (V*).

All further results concerning the existence of sensitive-discount optimal policies which are known to the authors (except for the papers of HORDIJK and SLADKY [58] and WIJNGAARD [122]) refer to the case N = -1 and were derived in connection with the HOWARD-algorithm for average maximization (see 2.4). The approach of WIJNGAARD [122] consists in requiring such conditions which are generally not fulfilled in our model if the state space is infinite. All the other publications deal with Markov chains having exactly one class of recurrent states. Therefore, in some of them, it turned out to be sufficient to suppose recurrence conditions weaker than (V**). E.g., in [58], the authors noted that there is a decision function f_N^* with $\mathfrak{G}_{N+1}(f_N^*) = \emptyset$ for every $N \in \mathbb{N}_{-1}$ if there are a state $x_0 \in E$, a $n_0 \in \mathbb{N}$, and a constant $\alpha > 0$ satisfying

$$P(f)^{n_0}(x,x_0) \geqq \alpha$$

for every $x \in E$, $f \in \mathbb{D}$.

3. STATE SPACE CLUSTERING IN MDP's

3.1. TRANSFORMATION OF THE MODEL

$\mathfrak{C}$-COMPATIBLE POLICIES

Consider an MDP $\mathfrak{M} = (E, \mathfrak{D}, D, p, r)$ with a countable state space E, a set E' with card $E' \leqq$ card E, and a surjective mapping $\varphi: E \to E'$. Further let us denote

$$\mathfrak{C} := \varphi^{-1}(\mathfrak{P}(E')),$$

the sub-σ-algebra of $\mathfrak{P}(E)$ generated by φ. In the same manner as in chapter 2 let $(X_n) = (X_n)_{n\in\mathbb{N}_0}$ be the observation process connected with the given MDM $\mathfrak{M}$ and $(Y_n) = (Y_n)_{n\in\mathbb{N}_0}$ the image (or cluster) process defined by $Y_n := \varphi(X_n)$, $n \in \mathbb{N}_0$, on the basic space $(A, \mathfrak{A})$.

Definition 3.1. Let $\pi = (f_0, f_1, \ldots) \in \mathfrak{R}$. π is called $\mathfrak{C}$-compatible if the relation $\varphi(x_k) = \varphi(x_k')$, $k = 0,\ldots,n \Rightarrow f_n(y_n) = f_n(y_n')$ is fulfilled for every $n \in \mathbb{N}_0$, all $y_n = (x_0,\ldots,x_n) \in E^{n+1}$, and all $y_n' = (x_0',\ldots,x_n') \in E^{n+1}$.
The subsets of all $\mathfrak{C}$-compatible policies of $\mathfrak{R}$, $\mathfrak{R}_M$, and $\mathfrak{R}_S$ are denoted by $\mathfrak{R}$, $\mathfrak{R}_M$, and $\mathfrak{R}_S$, respectively.

The $\mathfrak{C}$-compatibility of a policy can be interpreted as follows: If $(x_0,\ldots,x_n)$ and $(x_0',\ldots,x_n')$ are two state sequences of the original process which cannot be distinguished from each other by observing the image process, then we choose the same decision for these two sequences of states. This is reasonable since, in general, it is impossible to determine the corresponding state sequences of the original process by the exclusive observation of the state sequences of the image process.

Definition 3.2. A decision function $f \in \mathbb{D}$ is said to be $\mathfrak{C}$-compatible if it fulfils the condition

$$\varphi(x) = \varphi(y) \Longrightarrow f(x) = f(y) \qquad (3.1)$$

for $x,y \in E$.

The set of all $\mathfrak{C}$-compatible decision functions will be denoted by $\mathbb{D}^{\mathfrak{C}}$.

The preceding definition implies that a Markovian policy $(f_o, f_1, \ldots)$ is $\mathfrak{C}$-compatible, if and only if each of the decision functions $f_o, f_1, \ldots$ is a $\mathfrak{C}$-compatible one.

Note: Obviously, a decision function $f \in \mathbb{D}$ is $\mathfrak{C}$-compatible, if and only if it is constant on each of the sets $\varphi^{-1}(s)$, $s \in E'$. Likewise, every policy $\pi = (f_o, f_1, \ldots)$ is $\mathfrak{C}$-compatible, if and only if the functions f_n are constant on all subsets $\varphi^{-1}(s_o) \times \ldots \times \varphi^{-1}(s_n)$ for each $n \in \mathbb{N}_0$ and $s_o, \ldots, s_n \in E'$. Hence, we shall also use the notation "piece-wise constant" instead of "$\mathfrak{C}$-compatible" if this does not cause any confusion.

In the following, the function D will satisfy the following condition:

$$\varphi(x) = \varphi(y) \Rightarrow D(x) = D(y) \qquad (3.2)$$

for $x, y \in E$. Then we can define a decision field on the image (cluster) space E':
Let $\mathfrak{D}' := \mathfrak{D}$ and let

$$D'(y) := D(x) \text{ if } \varphi(x) = y$$

for $y \in E'$. Based on (3.2) this definition is independent of the choice of x and, in this way, it is a correct one. The triple $(E', \mathfrak{D}', D')$ is a decision field. The classes of policies defined here will be denoted by $\mathfrak{R}'$, $\mathfrak{R}'_M$, and $\mathfrak{R}'_S$. Further let $\mathbb{D}'$ be the set of all decision functions defined on E'. There is a close connection between the policies on $(E', \mathfrak{D}', D')$ and those of $(E, \mathfrak{D}, D)$ as we shall show in the following.

For the purpose let $\mathfrak{M}_n$ be the set of all functions $f.\ E^{n+1} \to \mathfrak{D}$ with $f(x_o, \ldots, x_n) \in D(x_n)$ for all $(x_o, \ldots, x_n) \in E^{n+1}$. We define $\mathfrak{M}_n'$ analogously. Moreover, let $\mathfrak{M}_n^{\mathfrak{C}}$ be the subset of all $\mathfrak{C}$-compatible functions of $\mathfrak{M}_n$. We define

$$\tilde{\theta}_n f := f \circ \psi^{n+1}$$

for every $f \in \mathfrak{M}_n'$, i.e.,

$$(\tilde{\theta}_n f)(x_o, \ldots, x_n) = f(\varphi(x_o), \ldots, \varphi(x_n))$$

holds for all $x_o, \ldots, x_n \in E$. We then obtain

Lemma 3.1. The mapping $\tilde{\theta}_n : f \to \tilde{\theta}_n f$ is a bijection from $\mathfrak{M}_n'$ onto $\mathfrak{M}_n^{\mathfrak{C}}$.

Proof: Due to (3.2) the relation

$$f(\varphi(x_o), \ldots, \varphi(x_n)) \in D'(\varphi(x_n))$$

is valid. Hence, we obtain for every $f \in \mathfrak{M}_n'$ that

$$\tilde{\theta}_n f \in \mathfrak{M}_n^{\mathfrak{C}}.$$

If $\tilde{\theta}_n f = \tilde{\theta}_n g$ with $f, g \in \mathfrak{M}_n'$, then

$$f(\varphi(x_o),\ldots,\varphi(x_n)) = g(\varphi(x_o),\ldots,\varphi(x_n))$$

for all $x_o,\ldots,x_n \in E$. Since φ is surjective it results that $f = g$, i.e., $\tilde{\theta}_n$ is injective. Finally, we choose any $f \in \mathfrak{M}_n^{\mathfrak{C}}$. Then we define a function on $(E')^{n+1}$ by

$$f_\varphi(y_o,\ldots,y_n) := f(x_o,\ldots,x_n)$$

for $\varphi(x_o) = y_o,\ldots,\varphi(x_n) = y_n$. This definition is correct because f is $\mathfrak{C}$-compatible. Now it holds that

$$f_\varphi(y_o,\ldots,y_n) \in D(x_n) = D'(y_n),$$

and, consequently, $f_\varphi \in \mathfrak{M}_n'$. Thus we obtain

$$\tilde{\theta}_n f = f_\varphi \circ \varphi^{n+1} = f,$$

i.e., $\tilde{\theta}_n$ is surjective.

Note: The mapping $f \to f_\varphi$ exactly describes the inverse mapping $\tilde{\theta}_n^{-1}: \mathfrak{M}_n^{\mathfrak{C}} \to \mathfrak{M}_n'$ of $\tilde{\theta}_n$ as the proof of the lemma 3.1 shows.

If we put $\theta f := \tilde{\theta}_o f = f_\varphi$ ($f \in \mathbb{D}'$), then we obtain the following result as a special case of lemma 3.1.

Lemma 3.2. The mapping $\theta: f \to \theta f$ is a bijection from $\mathbb{D}'$ onto $\mathbb{D}^{\mathfrak{C}}$.

Moreover, lemma 3.1 implies

Lemma 3.3. Let

$$\hat{\theta}\pi := (\tilde{\theta}_o f_o, \tilde{\theta}_1 f_1, \ldots)$$

for every policy $\pi \in \mathfrak{R}'$ of the form $\pi = (f_o, f_1, \ldots)$. Then $\hat{\theta}$ is a bijection from $\mathfrak{R}'$ onto $\mathfrak{R}^{\mathfrak{C}}$ with

$$\hat{\theta}\mathfrak{R}_M' = \mathfrak{R}_M^{\mathfrak{C}} \quad \text{and} \quad \hat{\theta}\mathfrak{R}_S' = \mathfrak{R}_S^{\mathfrak{C}} .$$

Proof: The fact that $\hat{\theta}$ bijectively maps $\mathfrak{R}'$ on $\mathfrak{R}^{\mathfrak{C}}$ immediately results from the corresponding properties of the mappings $\tilde{\theta}_n$, $n \in \mathbb{N}$, given by lemma 3.1. In order to prove the second assertion of lemma 3.3 let us remark that the function belongs to $\mathfrak{M}_n'$ if $\pi \in \mathfrak{R}_M'$. This implies

$$f_n(y_o,\ldots,y_n) = \tilde{f}_n(y_n)$$

with a suitable $f_n \in D'$ for such functions f_n. Obviously,

$$\tilde{\theta}_n f_n(x_o,\ldots,x_n) = \theta\tilde{f}_n(x_n)$$

for all $x_o,\ldots,x_n \in E$. Thus the assertion is proved in connection with lemma 3.2. ∎

The proposition of lemma 3.3 is of fundamental importance for the application of state space transformations in the investigation of Markov Decision Processes. Namely, if we succeeded in completing the decision field $(E', \mathfrak{D}', D')$ to a MDM $\mathfrak{M}' = (E', \mathfrak{D}', D', \bar{p}, \bar{r})$ by means of suitable functions $\bar{p}$ and $\bar{r}$ such that $\bar{p}$ and $\bar{r}$ correspond to p and r (in a certain sense), then we could expect that the solution of the decision problem with respect to the class $\mathfrak{R}^{\mathfrak{C}}$ is reducible to the solution of "ordinary" decision problems within the model $\mathfrak{M}'$. Taking into account the smaller "size" of the state space of $\mathfrak{M}'$ we can expect the problem to become easier. On the other hand, it seems to be obvious that we can expect results being at least as good as those which can be attained by restriction to the class $\mathfrak{R}^{\mathfrak{C}}$ (or sub-classes for it) for the original model, if calculations are based on the model $\mathfrak{M}'$ instead of $\mathfrak{M}$ with respect to one of the optimality criteria introduced in 2.1. It is reasonable that $\mathfrak{M}'$ should be constructed in such a way that the process $(Y_n) = (\varphi(X_n))$ can be interpreted as the basic observation process of $\mathfrak{M}'$. If we apply a stationary policy f^{∞} $(f \in \mathbb{D}')$ within such a model $\mathfrak{M}'$, then, according to the considerations in 2.1, (Y_n) must be a homogeneous Markov chain independent of the underlying initial distribution on E. Moreover, the "transition mechanism" has to be independent of the chosen initial distribution. This requires the process (X_n), i.e., its transition matrix, to be lumpable with respect to each piecewise constant stationary policy.

STRONGLY LUMPABLE MARKOV DECISION MODELS

In this section we shall show that the decision field $(E', \mathfrak{D}', D')$ can be provided with a transition law $\bar{p}$ and a pay-off-structure $\bar{r}$ if the original model is "strongly lumpable".

Definition 3.3. Let $\mathfrak{M} = (E, \mathfrak{D}, p, r)$ be an MDM. $\mathfrak{M}$ is said to be strongly lumpable if the following conditions are fulfilled:

(i) $x, y \in E, \quad \varphi(x) = \varphi(y) \Rightarrow D(x) = D(y)$, (3.2)

(ii) $f \in \mathbb{D} \Rightarrow r(f) \in B(E, \mathfrak{C})$, and (3.3)

(iii) $f \in \mathbb{D} \Rightarrow P(f)B(E, \mathfrak{C}) \subset B(E, \mathfrak{C})$. (3.4)

This definition expresses the following fact:
An MDM $\mathfrak{M}$ is strongly lumpable if the function D satisfies the compatibility condition (3.2), if for each piecewise constant decision function f the related pay-off-function r(f) is piecewise constant, and if the transition matrix P(f) is strongly lumpable (in the sense of definition 1.11). This property can be checked in the following manner:

Lemma 3.4. The MDM $\mathfrak{M} = (E, \mathfrak{D}, D, p, r)$ is strongly lumpable, if and only if it has the property that

(i) $D(x) = D(y)$,

(ii) $d \in D(x) \Rightarrow r(x,d) = r(y,d)$, and

(iii) $d \in D(x) \Rightarrow p(x,d,\varphi^{-1}(s)) = p(y,d,\varphi^{-1}(s))$ $(s \in E')$

for all $x,y \in E$ with $\varphi(x) = \varphi(y)$.

Proof: The condition (i) corresponds to the condition (i) of the definition 3.2. Let f be an element of $D^{\mathfrak{C}}$, i.e., f is constant on each of the sets $\varphi^{-1}(s)$, $s \in E'$. Further, let $f(x) =: f'(s)$ for an arbitrary $x \in \varphi^{-1}(s)$. Then $r(x,f(x)) = r(x,f'(s))$ is independent of $x \in \varphi^{-1}(s)$ for all $s \in E'$ and, consequently, $r(f) \in B(E, \mathfrak{C})$. In view of the supposition (iii) the matrix $P(f)$ satisfies the relation

$$p(x,f'(x), \varphi^{-1}(s')) = p(y,f'(s), \varphi^{-1}(s'))$$

for all $s,s' \in E'$ and all $x,y \in \varphi^{-1}(s)$. Due to lemma 1.24 this relation is equivalent to $P(f)B(E, \mathfrak{C}) \subset B(E, \mathfrak{C})$. Hence, the conditions (i) - (iii) imply that $\mathfrak{M}$ is a strongly lumpable model. We obtain the inversion analogously. ∎

Note: In the frequent case of $D(x) \equiv D_0$ $(x \in E)$ the condition (iii) of lemma 3.4 is easily to verify. Namely, we only need to examine whether the transition matrix $P(f)$ allows a strong lumpation for every constant decision function $f \equiv d$, $d \in D_0$.

Henceforth, our considerations will be based on the general condition that the considered model $\mathfrak{M}$ is strongly lumpable. Then the decision field $(E', \mathfrak{D}', D')$ can be supplied with a pay-off-structure $\bar{r}$, which is defined as follows:
For $s \in E'$, $d \in D'(s)$ let $\bar{r}(s,d) := r(x,d)$ if $\varphi(x) = s$ holds. This definition is correct in view of lemma 3.4.

There exists a norm-preserving isomorphism

$$\Phi: B(E', \mathfrak{P}(E')) \to B(E, \mathfrak{C})$$

as it has been shown above (lemma 1.22). Then we obtain

Lemma 3.5. $\Phi\bar{r}(f) = r(\theta f)$ for all $f \in D'$. (3.5)

Proof: The equations

$$\begin{aligned}\Phi\bar{r}(f) &= \bar{r}(f) \circ \varphi(x) \\ &= r(\varphi(x), f(\varphi(x))) \\ &= r(x,f(\varphi(x))) \\ &= r(f \circ \varphi)(x)\end{aligned}$$

are valid for all $x \in E$. This proves the assertion $\Phi\bar{r}(f) = r(\theta f)$ of the lemma. ∎

A transition law $\bar{p}$ on $K' \times E'$ is defined in a similar way as the pay-off-structure $\bar{r}$, where K' is given by

$$K' := \{(s,d): s \in E', d \in D'(s)\}.$$

Let

$$\bar{p}(s,d,s') := p(x,d, \varphi^{-1}(s'))$$

if $x \in \varphi^{-1}(s)$ for each pair $s,s' \in E$ and all $d \in D'(s)$. This definition is justified since the function $x \to p(x,d, \varphi^{-1}(s'))$ is constant because of the strong lumpability of $\mathfrak{M}$.

Moreover, we again refer to the section 1.4, whereas lemma 1.23 showed that the mapping $\tilde{\Phi}: \mathfrak{L}' \to \mathfrak{L}_{\mathfrak{C}}$ is a norm-preserving isomorphism.

Lemma 3.6. $(\tilde{\Phi}\bar{P})(f) = P(\theta f)_{\mathfrak{C}}$ for all $f \in \mathbb{D}'$.

Proof: The equation $\bar{P}(f) = \overline{P(\theta f)}$ holds in the sense of theorem 1.21. Hence the assertion of this lemma is a consequence of theorem 1.26. ∎

Note: The operator $(\tilde{\Phi}\bar{P})(f) = P(\theta f)_{\mathfrak{C}}$ on $B(E, \mathfrak{C})$ has always the matrix representation $P(\theta f)$, i.e.,

$$(\tilde{\Phi}\bar{P}(f))v = P(\theta f)_{\mathfrak{C}} v = P(\theta f)v, \qquad (3.6)$$

where $P(\theta f)$ can be interpreted as an operator on $B(E)$ as well as a transition matrix on E. Let us remark that the representation (3.6) of the operator $P(\theta f)_{\mathfrak{C}}$ by an $E \times E$-matrix need not be the only possible one.
Contrarily, the matrix $\bar{P}(f)$ and the operator $\bar{P}(f) = \Phi^{-1}P(\theta f)_{\mathfrak{C}}$ induced by $\bar{P}(f)$ correspond in a one-to-one manner for a given f.

Let τ be a policy with respect to the image process. The preceding assertions imply that both the transition matrices and the pay-off functions, which characterize the image process ruled by τ, closely correspond to the transition matrices and the pay-off functions of the original process ruled by $\hat{\theta}_{\tau}$. If we define the reward functions $\bar{V}_N^{\tau}$, $\bar{V}_{\beta}^{\tau}$ and $\bar{g}^{\tau}$ analogously to those of the original process, we can suppose that these reward functions and those of the original process also correspond in a very strictly sense. In the following this assumption will be confirmed.

REWARD CRITERIA IN A TRANSFORMED MODEL

If we want to define reward functions with respect to the MDM $\mathfrak{M}' = (E', \mathfrak{D}', D', \bar{p}, \bar{r})$ resulting from $\mathfrak{M} = (E, \mathfrak{D}, D, p, r)$ by lumpation, then the procedure of section 2.1 is possible to apply. We have to remark, however, that the space $(\bar{A}, \bar{\mathfrak{A}})$ with

$$\bar{A} = (E')^{\mathbb{N}_0} \text{ and } \bar{\mathfrak{A}} := \bigotimes_{\mathbb{N}_0} \mathfrak{P}(E')$$

plays the role of a basic space for such a construction, and that the sequence $(\bar{X}_n)_{n\in\mathbb{N}_0}$ of canonical projections $\bar{X}_n: \bar{A} \to E'$ can be interpreted as the basic process of the MDM $\mathfrak{M}'$. If $\tau \in \mathfrak{R}'$ is a policy of the form $\tau = (f_0, f_1, \ldots)$ and if $\mu \in \mathfrak{M}^1(E')$, then a probability measure on $(\bar{A}, \bar{\mathfrak{A}})$ is usually defined as follows:

$$\tilde{\mathbb{P}}_\mu^\tau(\bar{X}_0 = s_0, \ldots, \bar{X}_n = s_n) :=$$

$$\mu(s_0)\bar{p}(s_0, f_0(s_0), s_1) \ldots \bar{p}(s_{n-1}, f_{n-1}(s_0, \ldots, s_{n-1}) s_n). \qquad (3.7)$$

In order to avoid any difficulties arising from the use of different basic space we shall consider the measurable space $(A, \mathfrak{A})$ as the basic space. The sequence $(Y_n) = (\varphi(X_n))$ will be considered as the observation process of the model $\mathfrak{M}'$. Every decision problem within the model $\mathfrak{M}'$ can be interpreted in this way as a problem of optimal decision with respect to the image process $(Y_n) = (\varphi(X_n))$.

Let $\mathfrak{A}_{\mathfrak{C}} := \bigotimes_{\mathbb{N}_0} \mathfrak{C}$. We define a probability measure $\bar{\mathbb{P}}_\mu^\tau$ on $(A, \mathfrak{A}_{\mathfrak{C}})$:
Let

$$\bar{\mathbb{P}}_\mu^\tau(Y_0 = s_0, \ldots, Y_n = s_n) :=$$
$$\mu(s_0)\bar{p}(s_0, f_0(s_0), s_1) \ldots \bar{p}(s_{n-1}, f_{n-1}(s_0, \ldots, s_{n-1}), s_n) \qquad (3.8)$$

for each cylinder set

$$[Y_0 = s_0, \ldots, Y_n = s_n] = [X_0 \in \varphi^{-1}(s_0), \ldots, X_n \in \varphi^{-1}(s_n)],$$

$n \in \mathbb{N}_0$; $s_0, \ldots, s_n \in E'$. Obviously, the σ-algebra $\mathfrak{A}_{\mathfrak{C}}$ is generated by the system of such cylinder sets. Due to the extension theorem (theorem 5.7 [4]) $\bar{\mathbb{P}}_\mu^\tau$ can be uniquely continued to a probability measure on $\mathfrak{A}_{\mathfrak{C}}$. This measure will also be denoted by $\bar{\mathbb{P}}_\mu^\tau$. Comparing (3.7) and (3.8) we see that $(\bar{X}_n)$ on $(\bar{A}, \bar{\mathfrak{A}}, \tilde{\mathbb{P}}_\mu^\tau)$ and (Y_n) on $(A, \mathfrak{A}, \bar{\mathbb{P}}_\mu^\tau)$ are equivalent processes, i.e., their finite dimensional distributions coincide.

In the following we aim at describing the relation between the measures $\bar{\mathbb{P}}_\mu^\tau$ on $(A, \mathfrak{A}_{\mathfrak{C}})$ and $\mathbb{P}_{\bar{\mu}}^{\hat{\theta}\tau}$, which is to construct for the original model, where the initial distribution is still to be determined.

<u>Theorem 3.7.</u> Let $\bar{\mu} \in \mathfrak{M}^1(E)$ be an initial distribution with $\varphi\bar{\mu} = \mu$. Then

$$\bar{\mathbb{P}}_\mu^\tau = \mathbb{P}_{\bar{\mu}}^{\hat{\theta}\tau}|_{\mathfrak{A}_{\mathfrak{C}}}.$$

<u>Proof:</u> Firstly, we shall prove the identity of $\bar{\mathbb{P}}_\mu^\tau$ and $\mathbb{P}_{\bar{\mu}}^{\hat{\theta}\tau}$ on the system of cylinder sets of $\mathfrak{A}_{\mathfrak{C}}$ of the following form

$$[Y_0 = s_0, \ldots, Y_n = s_n] = \{\omega: Y_0(\omega) = s_0, \ldots, Y_n(\omega) = s_n\},$$

$n \in \mathbb{N}_0$; $s_0, \ldots, s_n \in E'$. For such a set we obtain the relations:

$$\mathbb{P}_{\bar{\mu}}^{\hat{\theta}\tau}(Y_0 = s_0, \ldots, Y_n = s_n) = \mathbb{P}_{\bar{\mu}}^{\hat{\theta}\tau}(X_0 \in \varphi^{-1}(s_0), \ldots, X_n \in \varphi^{-1}(s_n)) =$$

$$= \sum_{x_0\in\varphi^{-1}(s_0)} \cdots \sum_{x_n\in\varphi^{-1}(s_n)} \bar{\mu}(x_0)p(x_0,\theta_0 f_0(x_0),x_1)\cdots$$
$$\cdots p(x_{n-1},\theta_{n-1}f_{n-1}(x_0,\ldots,x_{n-1}),x_n)$$
$$= \sum_{x_0\in\varphi^{-1}(s_0)} \cdots \sum_{x_n\in\varphi^{-1}(s_n)} \bar{\mu}(x_0)p(x_0,f_0(\varphi(x_0)),x_1)\cdots$$
$$\cdots p(x_{n-1},f_{n-1}(\varphi(x_0),\ldots,\varphi(x_{n-1})),x_n)$$
$$= \Big(\sum_{x_0\in\varphi^{-1}(s_0)} \cdots \sum_{x_{n-1}\in\varphi^{-1}(s_{n-1})} \bar{\mu}(x_0)p(x_0,f_0(\varphi(x_0)),x_1)\cdots$$
$$\cdots p(x_{n-2},f_{n-2}(\varphi(x_0),\ldots,\varphi(x_{n-2})),x_{n-1})\Big)\bar{p}(s_{n-1},f_{n-1}(s_0,\ldots,s_{n-1})s_n),$$

$n \geq 2$, because $\mathfrak{M}$ is strongly lumpable and the function

$$x_{n-1} \to p(x_{n-1},f_{n-1}(\varphi(x_0),\ldots,\varphi(x_{n-1})),\varphi^{-1}(s_n))$$

is constant on $\varphi^{-1}(s_{n-1})$ (cf. lemma 3.4 (iii)). By a repeated application of this conclusion and

$$\mu(s_0) = \sum_{x_0\in\varphi^{-1}(s_0)} \bar{\mu}(x_0)$$

one obtains

$$\mathbb{P}^{\hat{\theta}\tau}_{\bar{\mu}}(Y_0 = s_0,\ldots,Y_n = s_n)$$
$$= \overline{\mu}(s_0)\bar{p}(s_0,f_0(s_0),s_1)\cdots\bar{p}(s_{n-1},f_{n-1}(s_0,\ldots,s_{n-1}),s_n)$$
$$= \overline{\mathbb{P}}^{\tau}_{\mu}(Y_0 = s_0,\ldots, Y_n = s_n).$$

The assertion then results from a theorem concerning the unique extension of probability measures (cf. [4]).

Using the probability measure $\overline{\mathbb{P}}^{\tau}_{\mu}$, which is associated with a policy $\tau \in \mathfrak{R}'$, we can define $\bar{V}^{\tau}_{N}$, $\bar{V}^{\tau}_{\beta}$, and $\bar{g}^{\tau}$ completely analogous to the functions V^{π}_{n}, V^{π}_{β}, and g^{π} of the original process, i.e.,

$$\bar{V}^{\tau}_{N}(s) = \overline{\mathbb{E}}^{\tau}_{s} \sum_{n=0}^{N} r(Y_n,f_n(Y_0,\ldots,Y_n)),$$

$$\bar{V}^{\tau}_{\beta}(s) = \overline{\mathbb{E}}^{\tau}_{s} \sum_{n=0}^{\infty} \beta^n \bar{r}(Y_n,f_n(Y_0,\ldots,Y_n)),$$

and

$$\bar{g}^{\tau}(s) = \varliminf_{N\to\infty} \frac{1}{N} \sum_{n=0}^{N-1} \overline{\mathbb{E}}^{\tau}_{s}\, \bar{r}(Y_n,f_n(Y_0,\ldots,Y_n))$$

(where $\overline{\mathbb{P}}^{\tau}_{s} := \overline{\mathbb{P}}^{\tau}_{\varepsilon_s}$, and $\overline{\mathbb{E}}^{\tau}_{s}(.)$ represents the mathematical expectation with respect to $\overline{\mathbb{P}}^{\tau}_{s}$ ($s \in E'$)). ∎

From theorem 3.7 we obtain the following

Corollary 3.8. Let $\bar{\mu} \in \mathfrak{M}^1(E)$ with $\varphi\bar{\mu} = \mu$.
Then

$$\bar{\mathbb{E}}^{\tau}_{\mu}\ \bar{r}(Y_n, f_n(Y_o,\ldots,Y_n)) = \mathbb{E}^{\hat{\theta}\tau}_{\bar{\mu}} r(X_n, \theta_n f_n(X_o,\ldots,X_n)) \tag{3.9}$$

for all $n \in \mathbb{N}_0$.

Proof: The function $\bar{r}(Y_n, f_n(Y_o,\ldots,Y_n))$ is $\mathfrak{A}_{\mathfrak{C}}$-measurable. We obtain

$$\bar{\mathbb{E}}^{\tau}_{\mu}\bar{r}(Y_n, f_n(Y_o,\ldots,Y_n)) = \int_A \bar{r}(\varphi(X_n), f_n(\varphi(X_o),\ldots,\varphi(X_n)))d\bar{P}^{\tau}_{\mu}$$

$$= \int_A \bar{r}(\varphi(X_n), f_n(\varphi(X_o),\ldots,\varphi(X_n)))dP^{\hat{\theta}\tau}_{\bar{\mu}}$$

$$= \int_A r(X_n, \tilde{\theta}_n f_n(X_o,\ldots,X_n))dP^{\hat{\theta}\tau}_{\bar{\mu}} = \mathbb{E}^{\hat{\theta}\tau}_{\bar{\mu}}\ r(X_n, \tilde{\theta}_n f_n(X_o,\ldots,X_n))$$

by theorem 3.7. ∎

Note: In particular, we obtain the following proposition:

$$\bar{\mathbb{E}}^{\tau}_{\varphi(x)}\bar{r}(Y_n, f_n(Y_o,\ldots,Y_n)) = \mathbb{E}^{\hat{\theta}\tau}_{x} r(X_n, \tilde{\theta}_n f_n(X_o,\ldots,X_n))$$

for any $x \in E$.
If $\tau \in \mathfrak{R}'_M$ with $\tau = (f_o, f_1, \ldots)$, then the equation

$$\bar{P}(f_o)\bar{P}(f_1)\ldots\bar{P}(f_n)\bar{r}(f_n)(\varphi(x)) = P(\theta f_o)P(\theta f_1)\ldots P(\theta f_n)r(\theta f_n)(x)$$

follows. This equation is also available from lemmas 3.5 and 3.6.

The main result of this section is the following

Theorem 3.9. Let $\mathfrak{M} = (E, \mathfrak{D}, D, p, r)$ be strongly lumpable. Then for each policy $\tau \in \mathfrak{R}'$ the following relations hold:

(i) $\bar{V}^{\tau}_N \circ \varphi = V^{\hat{\theta}\tau}_N$, (3.10)

(ii) $\bar{V}^{\tau}_{\beta} \circ \varphi = V^{\hat{\theta}\tau}_{\beta}$, and (3.11)

(iii) $\bar{g}^{\tau} \circ \varphi = g^{\hat{\theta}\tau}$. (3.12)

Proof: Due to the theorem 3.7 and (3.9) we obtain the equation:

$$\sum_{n=0}^{N} \bar{\mathbb{E}}^{\tau}_{\varphi(x)}\bar{r}(Y_n, f_n(Y_o,\ldots,Y_n)) = \sum_{n=0}^{N} \mathbb{E}^{\hat{\theta}\tau}_{x}\ r(X_n, \tilde{\theta}_n f_n(X_o,\ldots,X_n)). \quad (*)$$

This immediately implies (i) and (iii). On the basis of the supposition

$$|\bar{r}(s,d)| \leq \tilde{k} < \infty$$

for all $s \in E'$, $d \in D'(s)$, we apply the Lebesgue Theorem of dominated convergence and obtain:

$$\bar{V}^{\tau}_{\beta}\circ\varphi(x) = \bar{\mathbb{E}}^{\tau}_{\varphi(x)}\sum_{n=0}^{\infty} \beta^n \bar{r}(Y_n, f_n(Y_o,\ldots,Y_n)) =$$

$$= \lim_{N\to\infty} \sum_{n=0}^{N} \bar{\mathbb{E}}^{\tau}_{\varphi(x)} \beta^n \bar{r}(Y_n, f_n(Y_0, \ldots, Y_n)).$$

Thus (ii) results from (*). ∎

Theorem 3.9 admits the following interpretation:
If the cluster process is governed by any policy $\tau \in \mathfrak{R}'$, then the reward function under consideration can be obtained by a simple transformation of the corresponding reward function of the original process if it is governed by $\hat{\theta}\tau$ corresponding to τ.
Reversely: If the original process is governed by a $\mathfrak{C}$-compatible (i.e., piecewise constant) policy $\pi \in \mathfrak{R}$, then the related reward function is again piecewise constant. Then its calculation can be replaced by the calculation of the corresponding reward function of the cluster process $\hat{\theta}^{-1}\pi$ (this, in general, will be advantageous).

Now, the comparison of two policies $\tau_1, \tau_2 \in \mathfrak{R}$ is possible if we use the following obvious

Corollary 3.10. Let $\mathfrak{M} = (E, \mathfrak{D}, D, p, r)$ be strongly lumpable. Then

(i) $\bar{V}_N^{\tau_1} \geq \bar{V}_N^{\tau_2} \Longleftrightarrow V_N^{\hat{\theta}\tau_1} \geq V_N^{\hat{\theta}\tau_2}$ $(N \in \mathbb{N}_0)$,

(ii) $\bar{V}_\beta^{\tau_1} \geq \bar{V}_\beta^{\tau_2} \Longleftrightarrow V_\beta^{\hat{\theta}\tau_1} \geq V_\beta^{\hat{\theta}\tau_2}$ $(\beta \in [0,1))$,

(iii) $\bar{g}^{\tau_1} \geq \bar{g}^{\tau_2} \Longleftrightarrow g^{\hat{\theta}\tau_1} \geq g^{\hat{\theta}\tau_2}$, and

(iv) $\underline{\lim}_{\rho\downarrow 0}\, \rho^{-n}(\bar{V}_\beta^{\tau_1} - \bar{V}_\beta^{\tau_2}) \geq 0 \Longleftrightarrow \underline{\lim}_{\rho\downarrow 0}\, \rho^{-n}(V_\beta^{\hat{\theta}\tau_1} - V_\beta^{\hat{\theta}\tau_2}) \geq 0.$

All these equivalence relations remain valid if "≥" is replaced by ">".

EXISTENCE AND CALCULATION OF OPTIMAL POLICIES FOR THE ORIGINAL AND THE TRANSFORMED MODEL - A GENERAL PROPOSITION

In the preceding section we considered the correspondence between the policies and the reward functions of the transformed model and piecewise constant policies and the related reward functions of the original model. Now we shall investigate the questions concerning the connections between the two models with respect to existence of optimal policies and their calculations.
An immediate consequence of the above considerations is:

Theorem 3.11. Let C be any subset of $\mathfrak{R}'$ and N a non-negative integer.
(i) A policy $\tau^* \in \mathfrak{R}'$ is N-step-optimal on C, if and only if $\hat{\theta}\tau^*$ is N-step-optimal on $\hat{\theta}C$.
(ii) There exists an N-step-optimal policy on C in $\mathfrak{R}'$, if and only if

there exists an N-step-optimal policy in $\mathfrak{R}$ which is piecewice constant on $\hat{\theta}C$.
(iii) The propositions (i) and (ii) remain valid if one replaces "N-step-optimal" by "β-optimal" ($\beta \in [0,1)$), "average-optimal" or "N-optimal" ($N \in \mathbb{N}_{-1}$), respectively, $\mathfrak{R}'$ by $\mathfrak{R}'_M$ or $\mathfrak{R}'_S$, and $\mathfrak{R}$ by $\mathfrak{R}_M$ or $\mathfrak{R}_S$, respectively.

Proof:
(i) As defined above a policy τ^* is N-step-optimal on C if the inequality $\overline{V}_N^{\tau^*} \geq \overline{V}_N^{\tau}$ is fulfilled for all $\tau \in C$. Corollary 3.10 shows that this is equivalent to $V_N^{\hat{\theta}\tau^*} \geq V_N^{\hat{\theta}\tau}$ for all $\tau \in C$ and to $V_N^{\hat{\theta}\tau^*} \geq V_N^{\pi}$ for all $\pi \in \hat{\theta}C$. Therefore $\hat{\theta}\tau^*$ is N-step-optimal on $\hat{\theta}C$. The inversion can be proved in an analogous way.
(ii) This proposition is an immediate consequence of (i).
(iii) Taking into account the relations

$$\hat{\theta}\mathfrak{R}'_M = \mathfrak{R}_M^{\mathbb{C}} \text{ and } \hat{\theta}\mathfrak{R}'_S = \mathfrak{R}_S^{\mathbb{C}}$$

(cf. lemma 3.3) we obtain the propositions in a similar way as above.∎

The theorem 3.11 shows that any optimality assertion within the transformed model can be formulated as an optimality assertion with respect to the class of piecewise constant policies within the original model, where the optimal policies are calculable from another by the (simple) transformation $\hat{\theta}$.
If the optimality of a policy is connected with the validity of any form of a BELLMAN equation, the assertion of theorem 3.11 can be sharpened: The exclusive consideration of piecewise constant policies of the original model does not mean any restriction. We shall thoroughly make this clear in connection with assertions concerning the calculations of optimal policies and policy improvements, respective.

3.2. FINITE HORIZON AND DISCOUNTED MDP's

PROBLEMS WITH FINITE HORIZON

Lemma 3.12. Let $v \in B(E, \mathbb{C})$ be a real function. Then

$$\max_{f \in \mathbb{D}} \{r(f) + P(f)v\} = \max_{f \in \mathbb{D}^{\mathbb{C}}} \{r(f) + P(f)v\}.$$

Proof: The maximum

$$\max_{f \in \mathbb{D}} \{r(f) + P(f)v\}$$

exists and it is attained by each decision function $f^* \in \mathbb{D}$ with the property

$$r(x,f^*(x)) + \sum_{y\in E} p(x,f^*(x),y)v(y) =$$

$$= \max_{d\in D(x)} \{r(x,d) + \sum_{y\in E} p(x,dy)v(y)\},$$

$x \in E$. Let f^* be such a decision function, $\{x_0,x_1,\ldots\}$ a complete system of representatives of all classes $\varphi^{-1}(s)$, $s \in E'$, and let us denote $\bar{f}(x) := f^*(x_n)$ if $\varphi(x) = \varphi(x_n)$ for all $x \in E$.
Obviously, $\bar{f} \in \mathbb{D}^{\mathfrak{C}}$ and

$$r(f^*) + P(f^*)v \geq r(\bar{f}) + P(\bar{f})v.$$

Now let us assume that

$$[r(f^*) + P(f^*)v](x^*) > [r(\bar{f}) + P(\bar{f})](x^*) \qquad (^{o})$$

is fulfilled for any $x^* \in E$. There is no loss of generality if we assume that $\varphi(x^*) = \varphi(x_0)$. In view of the strong lumpability of $\mathfrak{M}$ and the $\mathfrak{C}$-measurability of v we obtain the identities

$$r(x_0,f^*(x^*)) + \sum_{y\in E} p(x_0,f^*(x^*),y)v(y) =$$

$$= r(x^*,f^*(x^*)) + \sum_{y\in E} p(x^*,f^*(x^*),y)v(y)$$

and, according to the definition of $\bar{f}$,

$$r(x^*,\bar{f}(x^*)) + \sum_{y\in E} p(x^*,\bar{f}(x^*),y)v(y) =$$

$$= r(x_0,f(x^*)) + \sum_{y\in E} p(x_0,\bar{f}(x^*),y)v(y)$$

$$= r(x_0,f^*(x_0)) + \sum_{y\in E} p(x_0,f^*(x_0),y)v(y).$$

(o) implies the inequality:

$$r(x_0,f^*(x^*)) + \sum_{y\in E} p(x_0,f^*(x^*),y)v(y) >$$

$$> r(x_0,f^*(x_0)) + \sum_{y\in E} p(x_0,f^*(x_0),y)v(y),$$

which contradicts the definition of f^*. ∎

Let $\bar{V}^*_N(s) := \sup_{\tau\in\mathfrak{R}'} \bar{V}^{\tau}_N(s)$ and let $\bar{V}^*_N$ be the function defined by this relation.

<u>Theorem 3.13.</u> Let $N \in \mathbb{N}_0$. Then

$$\bar{V}^*_N \circ \varphi = \bar{V}^*_N ,$$

and there is an N-step-optimal policy τ^*_N on $\mathfrak{R}'$ in $\mathfrak{R}'_M$. The policy $\hat{\theta}\tau^*_N$ is an N-step-optimal policy on $\mathfrak{R}$ in $\mathfrak{R}^{\mathfrak{C}}_M$.

<u>Proof:</u> Due to the theorem 2.1

$$\bar{V}^*_o = \max_{f \in \mathbb{D}'} \bar{r}(f) = \bar{r}(f_o)$$

and

$$\bar{V}^*_n = \max_{f \in \mathbb{D}'} \{\bar{r}(f) + \bar{P}(f)\bar{V}^*_{n-1}\} = \bar{r}(f_n) + \bar{P}(f_n)\bar{V}^*_{n-1}$$

for certain decision functions $f_o, f_1, \ldots$ and for $n \geq 1$. If f'_{N+1}, f'_{N+2}, ... are any decision functions from $\mathbb{D}'$, then the policy

$$\tau^*_N := (f_N, f_{N-1}, \ldots, f_1, f_o, f'_{N+1}, f'_{N+2}, \ldots)$$

is N-step-optimal on $\mathcal{R}'$ and Markovian. The remaining part of the proposition will be proved inductively. Due to (3.5), (3.6), (3.10), and, additionally, due to lemma 3.4 (ii), theorem 2.1, and the monotony of Φ we obtain the equations

$$V_o^{\hat{\theta}\tau^*_o} = \Phi\bar{V}_o^{\tau^*_o} = \Phi\bar{V}^*_o = (\max_{f \in \mathbb{D}'} \bar{r}(f)) \circ \varphi = \max_{f \in \mathbb{D}'} (\bar{r}(f) \circ \varphi) =$$

$$= \max_{h \in \mathbb{D}^{\mathbb{C}}} r(h) = \max_{h \in \mathbb{D}} r(h) = V^*_o .$$

Let us assume that the equation

$$V_n^{\hat{\theta}\tau_n} = \bar{V}^*_n \circ \varphi = V^*_n \qquad (*)$$

holds for some $n \in \mathbb{N}_0$. Then, by analogous considerations (where we use lemma 3.12 instead of lemma 3.4 (ii)), we obtain the identities:

$$V_{n+1}^{\hat{\theta}\tau^*_{n+1}} = \Phi\bar{V}_{n+1}^{\tau^*_{n+1}} = \Phi\bar{V}^*_{n+1} = \max_{f \in \mathbb{D}'} \{\bar{r}(f) + \bar{P}(f)\bar{V}^*_n\} \circ \varphi$$

$$= \max_{f \in \mathbb{D}'} \{\bar{r}(f)\circ\varphi + (\tilde{\Phi}\bar{P}(f))(\bar{V}^*_n \circ \varphi)\} =$$

$$= \max_{f \in \mathbb{D}'} \{r(\theta f) + P(\theta f)V^*_n\} = \max_{h \in \mathbb{D}^{\mathbb{C}}} \{r(h) + P(h)V^*_n\}$$

$$= \max_{h \in \mathbb{D}} \{r(h) + P(h)V^*_n\} = V^*_{n+1}.$$

Hence, we obtain (*) for all $n \in \mathbb{N}_0$. The proof is complete, because the policy $\hat{\theta}\tau^*_n$ is an element of $\mathcal{R}^{\mathbb{C}}_M$ for every n. ∎

Thus the proof of the following corollary is obvious.

<u>Corollary 3.14.</u>

(i) The recursive relations

$$V^*_n = \max_{f \in \mathbb{D}^{\mathbb{C}}} \{r(f) + P(f)V^*_{n-1}\} \qquad (3.13)$$

with

$$V_0^* = \max_{f \in \mathbb{D}^{\mathfrak{C}}} r(f)$$

hold for $n \geq 1$.

(ii) Let $(f_0, f_1, \ldots)$ be any sequence from $\mathbb{D}^{\mathfrak{C}}$ with $V_0^* = r(f_0)$ and $V_n^* = r(f_n) + P(f_n)V_{n-1}^*$ for each $n \geq 1$. Then the sequence

$$\pi_n^* := (f_N, \ldots, f_1, f_0, f'_{N+1}, f'_{N+2}, \ldots),$$

$N \in \mathbb{N}_0$, is an N-step-optimal policy, where f'_{N+1}, f'_{N+2}, ... are arbitrary decision functions.

<u>Notes:</u>

1. Let us emphasize that the identity (3.13), in general, does not hold in the case of models which are not strongly lumpable. Beyond it, the expression

$$\max_{f \in \mathbb{D}^{\mathfrak{C}}} \{r(f) + P(f)V_{n-1}^*\}$$

need not be defined. Appropriate examples can easily be specified.

2. The special feature of (3.13) can be formulated (with the notations of the remark 2.1) in the following way. Let $s \in E'$ and x_s be any representative of the class $\varphi^{-1}(s)$. Then for all $x \in \varphi^{-1}(s)$ and $n \in \mathbb{N}_0$

$$M_n(x) = M_n(x_s).$$

This implies

$$\bigcap_{x \in \varphi^{-1}(s)} M_n(x) = \emptyset. \quad \rfloor$$

The propositions concerning N-step problems which have been described till now can be summarized as follows: If the model $\mathfrak{M}$ is strongly lumpable, then the N-step-optimal reward function is $\mathfrak{C}$-measurable for every $N \in \mathbb{N}_0$ and can be attained by a $\mathfrak{C}$-compatible policy. Not only the optimal reward function but also the related $\mathfrak{C}$-compatible policies can be obtained by using the mapping φ of the optimal reward function and of the related policies with respect to the transformed model. This proposition is also reversible. In particular, in order to solve the original problem it is sufficient to use the transformed model, which is reduced in its dimension. If we apply the feed-back technique within the original model we can restrict ourselves to the solution of (3.13) for a complete system of representatives x_s, $s \in E$.

<u>THE DISCOUNTED TOTAL REWARD</u>

Let

$$\bar{V}_\beta^*(s) := \sup_{\tau \in \mathfrak{R}'} \bar{V}_\beta^\tau(s) \qquad (\beta \in [0,1)).$$

Theorem 3.15. There exists a β-discounted-optimal stationary policy $\tau_\beta = f_\beta^\infty$ on $\mathcal{R}'$ for every $\beta \in [0,1)$. Moreover,

$$\bar{V}^*_\beta \circ \varphi = \bar{V}_\beta^{\tau_\beta} \circ \varphi = V_\beta^{\hat{\theta}\tau_\beta} = V_{\beta *} ,$$

i.e., the $\mathfrak{C}$-compatible stationary policy $\hat{\theta}\tau_\beta$ is β-discounted-optimal on $\mathcal{R}$.

Proof: Due to the proposition of the theorem 2.4 we obtain the equations

$$\bar{V}^*_\beta = \max_{f \in D'} \{\bar{r}(f) + \beta\bar{P}(f)\bar{V}^*_\beta\}$$

$$= \bar{r}(f_\beta) + \bar{P}(f_\beta)\bar{V}^*_\beta$$

and

$$\bar{V}^*_\beta = \bar{V}_\beta^{\tau_\beta} \quad \text{with} \quad \tau_\beta := f_\beta^\infty .$$

Consequently, due to (3.11) and the monotony of Φ we obtain the relations:

$$V_\beta^{\hat{\theta}\tau_\beta} = \underline{\Phi}\,\bar{V}_\beta^{\tau_\beta} = \Phi\bar{V}^*_\beta = \max_{f \in D'} \{\bar{r}(f) + \beta\bar{P}(f)\bar{V}^*_\beta\} \circ \varphi$$

$$= \max_{f \in \mathbb{D}'} \{\bar{r}(f) \circ \varphi + \beta(\tilde{\Phi}\bar{P}(f))(\bar{V}^*_\beta \circ \varphi)\}.$$

(3.5), (3.6), lemma 3.3, and lemma 3.12 imply

$$V_\beta^{\hat{\theta}\tau_\beta} = \max_{f \in \mathbb{D}'} \{r(\theta f) + \beta P(\theta f)(\bar{V}^*_\beta \circ \varphi)\}$$

$$= \max_{h \in \mathbb{D}^{\mathfrak{C}}} \{r(h) + \beta P(h)(\bar{V}^*_\beta \circ \varphi)\}$$

$$= \max_{h \in \mathbb{D}} \{r(h) + \beta P(h)(\bar{V}^*_\beta \circ \varphi)\}.$$

Taking theorem 2.4 into account we finally obtain

$$V_\beta^{\hat{\theta}\tau_\beta} = \bar{V}^*_\beta \circ \varphi = V^*_\beta .$$

The policy $\hat{\theta}\tau_\beta$ is stationary according to lemma 3.3. ∎

Corollary 3.16. The optimal β-discounted total reward V^*_β satisfies the fundamental equation

$$V^*_\beta = \max_{f \in \mathbb{D}^{\mathfrak{C}}} \{r(f) + \beta P(f)V^*_\beta\},$$

and the equation

$$V^*_\beta = V_\beta^{f^*}$$

holds for every function $f^* \in \mathbb{D}^{\mathfrak{C}}$ with

$$V^*_\beta = r(f^*) + \beta P(f^*)V^*_\beta .$$

In the following we shall show that the application of the HOWARD algorithm to the transformed model is equivalent to the application of this algorithm to $\mathfrak{C}$-compatible policies with respect to the original model.

We define analogously to 2.2 for decision functions $f,g \in \mathbb{D}$:

$$\overline{\psi}_\beta(g,f) := \overline{r}(g) + \beta\overline{P}(g)\overline{V}^f_\beta - \overline{V}^f_\beta ,$$

$$\overline{\mathfrak{G}}_\beta(f) := \{g' \in \mathbb{D}' : \overline{\psi}_\beta(g',f) > 0\}, \text{ and}$$

$$\mathfrak{G}^{\mathfrak{C}}_\beta(h) := \{h' \in \mathbb{D}^{\mathfrak{C}} : \psi_\beta(h',h) > 0\}$$

for $h \in \mathbb{D}^{\mathfrak{C}}$.

<u>Theorem 3.17.</u> Let $f \in \mathbb{D}'$. Then there hold the following equivalence relations:

(i) $g \in \overline{\mathfrak{G}}_\beta(f) \Longleftrightarrow \theta g \in \mathfrak{G}^{\mathfrak{C}}_\beta(\theta f) \subset \mathfrak{G}_\beta(\theta f)$

and

(ii) $\mathfrak{G}^{\mathfrak{C}}_\beta(\theta f) = \emptyset \Longleftrightarrow \mathfrak{G}_\beta(\theta f) = \emptyset$.

<u>Proof:</u>

(i) Following (3.5), (3.6), (3.11), and in view of the monotony of Φ we see that the following propositions are equivalent:

$$g \in \mathfrak{G}_\beta(f),$$

$$\overline{r}(g) + \beta\overline{P}(g)\overline{V}^f_\beta - \overline{V}^f_\beta > 0,$$

$$\overline{r}(g)\circ\varphi + \beta(\tilde{\Phi}\overline{P}(g))(\overline{V}^f_\beta\circ\varphi) - \overline{V}^f_\beta\circ\varphi > 0,$$

$$r(\theta g) + \beta P(\theta g)V^{\theta f}_\beta - V^{\theta f}_\beta > 0 \text{ and}$$

$$\theta g \in \mathfrak{G}^{\mathfrak{C}}_\beta(\theta f).$$

(ii) As $\mathfrak{G}^{\mathfrak{C}}_\beta(\theta f) \subset \mathfrak{G}_\beta(\theta f)$ holds, it suffices to prove the relation

$$\mathfrak{G}^{\mathfrak{C}}_\beta(\theta f) = \emptyset \Rightarrow \mathfrak{G}_\beta(\theta f) = \emptyset.$$

Let us assume that $\mathfrak{G}_\beta(\theta f)$ is non-empty. Then there exist a state $x_o \in E$ and a decision function $g \in \mathbb{D}$ with $\psi_\beta(g, \theta f)(x_o) > 0$. We define a function $\overline{g}$ on E:

$$\overline{g}(x) := \begin{cases} g(x_o) & \text{for } \varphi(x) = \varphi(x_o) \\ f(x) & \text{otherwise.} \end{cases}$$

Since $\theta f \in \mathbb{D}^{\mathfrak{C}}$ we have $g \in \mathbb{D}^{\mathfrak{C}}$. Moreover,

$$\psi_\beta(\overline{g}, \theta f)(x) = 0$$

for $x \in E$ with $\varphi(x) = \varphi(x_o)$. Furthermore, we obtain

$$\begin{aligned}\psi_\beta(\bar{g},\theta f)(x) &= r(x,\bar{g}(x)) + \sum_{y\in E} p(x,\bar{g}(x),y)V_\beta^{\theta f}(y) - V_\beta^{\theta f}(x)\\ &= r(x,g(x_0)) + \sum_{y\in E} p(x,g(x_0),y)V_\beta^{\theta f}(y) - V_\beta^{\theta f}(x)\\ &= r(x_0,g(x_0)) + \sum_{y\in E} p(x_0,g(x_0),y)V_\beta^{\theta f}(y) - V_\beta^{\theta f}(x_0)\\ &= \psi_\beta(g,\theta f)(x_0) > 0\end{aligned}$$

due to the $\mathfrak{C}$-measurability of $V_\beta^{\theta f}$ and the strong lumpability of $\mathfrak{M}$ for any $x \in E$ with $\varphi(x) = \varphi(x_0)$. This implies that $\bar{g} \in \mathfrak{G}_\beta^{\mathfrak{C}}(\theta f)$. This contradicts the supposition that $\mathfrak{G}_\beta(\theta f) = \emptyset$. ∎

In particular, theorem 3.17 shows that we can restrict ourselves to the calculation of a new $\mathfrak{C}$-compatible "better" policy if we apply the HOWARD iteration procedure starting from a $\mathfrak{C}$-compatible decision function. That means, if $f \in \mathbb{D}^{\mathfrak{C}}$ is any iterative with $\mathfrak{G}_\beta(f) \neq \emptyset$, then also $\mathfrak{G}_\beta^{\mathfrak{C}} \neq \emptyset$. Practically, one does not choose any decision function belonging to $\mathfrak{G}_\beta(f)$ as a next iterative, but one looks for a function g with

$$\psi_\beta(g,f) = \max_{h\in\mathbb{D}} \psi_\beta(h,f).$$

The following lemma, whose proof can be given analogously to theorem 3.17 (ii), shows that we can choose such a function $g \in \mathbb{D}^{\mathfrak{C}}$.

Lemma 3.18. Let $f \in \mathbb{D}^{\mathfrak{C}}$. Then, for each $\beta \in [0,1)$,

$$\max_{h\in\mathbb{D}} \psi_\beta(h,f) = \max_{h\in\mathbb{D}^{\mathfrak{C}}} \psi_\beta(h,f).$$

Without going into detail let us remark that it is sufficient for us to operate the HOWARD algorithm on a complete system of representatives x_s, $s \in E'$, but not on the whole state space E, if one starts from a $\mathfrak{C}$-compatible decision function with respect to the original model. Likewise, it is sufficient to solve the problem with respect to the transformed model in order to find a β-optimal decision function of the original model. In this case the backward transformatives of the optimal discounted total reward and of each of the β-optimal decision functions are the optimal discounted total reward and the β-optimal decision functions of the original model, respectively.

3.3. AVERAGE REWARD OPTIMIZATION

The present section deals with the connection between the models $\mathfrak{M}$ and $\mathfrak{M}'$ with respect to the maximization of the average reward. Firstly, we are interested in the existence of average optimal policies.

Theorem 3.19. If there are a constant g^* and a bounded real function $v^* \in E'$ which satisfies the fundamental equation

$$g^* \mathbf{1} + v = \max_{f \in \mathbb{D}'} \{r(f) + P(f)v^*\}, \tag{3.14}$$

then:

(i) There exists an average optimal policy $\tau^* = (f^*)^\infty$ with $f^* \in \mathbb{D}'$ and

$$g^* \mathbf{1} + v^* = \bar{r}(f^*) + \bar{P}(f^*)v^*$$

is average optimal on $\mathcal{R}'$, where $g^* \mathbf{1} = \bar{g}^{\tau^*}$.

(ii) The constant g^* and the function $v^* \circ \varphi$ solve the fundamental equation

$$g^* \mathbf{1} + v^* \circ \varphi = \max_{h \in \mathbb{D}} \{r(h) + P(h)(v^* \circ \varphi)\},$$

and there exists an average optimal policy $\pi^* \in \mathcal{R}$ with $g^* = g^{\pi^*}$. Every stationary $\mathfrak{C}$-compatible policy $\theta\tau^*$ is average optimal on $\mathcal{R}$, where τ^* is defined as in (i).

Proof:

(i), see theorem 2.25.

(ii) In view of lemma 3.3, (3.5), and (3.6) the application of the monotone transformation Φ to both sides of (3.14) implies the identity

$$g^* \mathbf{1} + v^* \circ \varphi = \max_{h \in \mathbb{D}^{\mathfrak{C}}} \{r(h) + P(h)(v^* \circ \varphi)\}.$$

Since $v^* \circ \varphi \in B(E, \mathfrak{C})$ is real one obtains the equation

$$g^* 1 + v^* \circ \varphi = \max_{h \in \mathbb{D}} \{r(h) + P(h)(v^* \circ \varphi)\}.$$

Due to theorem 2.25 there is an average optimal policy on $\mathcal{R}$. The remaining part of the assertion follows from theorem 3.11. ∎

Note: Some conditions which guarantee the existence of a solution of (3.14) are given in 2.4.

Under the restrictive condition of a finite state space E we obtain the following result concerning the multichain case.

Theorem 3.20. Let E be a finite state space, and let g^* and v^* be real-valued functions on E' which satisfy the functional equation system

$$g^* = \max_{f \in \mathbb{D}'} \bar{P}(f) g^* \tag{3.15}$$

and

$$g^* + v^* = \max_{f \in \mathbb{D}'_{-1}} \{\bar{r}(f) + \bar{P}(f)v^*\} \tag{3.16}$$

with

$$\mathbb{D}'_{-1} := \{f^* \in \mathbb{D}' : g^* = \bar{P}(f^*)g^*\}.$$

Then:

(i) If $f^* \in \mathbb{D}'$ is a decision function fulfilling the conditions

$$g^* = \bar{P}(f^*)g^* \text{ and } g^* + v^* = \bar{r}(f^*) + \bar{P}(f^*)v^*,$$

then $\tau^* := (f^*)^\infty$ is average optimal on $\mathfrak{R}'$ with $g^* = \bar{g}^{\tau^*}$.

(ii) The functions $g^* \circ \varphi$ and $v^* \circ \varphi$ solve the equation system

$$g^* \circ \varphi = \max_{h \in \mathbb{D}} P(h)(g^* \circ \varphi) \tag{3.17}$$

and

$$g^* \circ \varphi + v^* \circ \varphi = \max_{h \in \mathbb{D}^*_{-1}} \{r(h) + P(h)(v^* \circ \varphi)\} \tag{3.18}$$

with

$$\mathbb{D}^*_{-1} = \{h^* \in \mathbb{D} : g^* \circ \varphi = P(h^*)(g^* \circ \varphi)\}.$$

Each policy $\hat{\theta}\tau^*$ is average optimal on $\mathfrak{R}$ and satisfies the equation,

$$g^* \circ \varphi = g^{\hat{\theta}\tau^*},$$

where τ^* is defined as above (i).

<u>Proof:</u> Part (i) follows from theorem 2.26.

(ii) At first, we obtain the identity

$$g^* \circ \varphi = \max_{h \in \mathbb{D}^{\mathbb{C}}} P(h)(g^* \circ \varphi)$$

from (3.15). In order to prove (3.17) it is sufficient to show that the equation

$$\max_{h \in \mathbb{D}^{\mathbb{C}}} P(h)(g^* \circ \varphi) = \max_{h \in \mathbb{D}} P(h)(g^* \circ \varphi) \tag{1*}$$

is valid. For this reason we denote

$$D^*_{-1}(x) := \{d \in D(x) : \sum_{y \in E} p(x,d,y)(g^* \circ \varphi)(y) =$$

$$= \max_{d' \in D(x)} \sum_{y \in E} p(x,d',y)(g^* \circ \varphi)(y)\}$$

for $x \in E$. Now let $x, y \in E$ be arbitrarily chosen such that $\varphi(x) = \varphi(y)$, and let $d \in D^*_{-1}(x)$, $d' \in D^*_{-1}(y)$. As defined above we obtain

$$\sum_{z \in E} p(y,d,z) g^* \circ \varphi(z) \leq \sum_{z \in E} p(y,d',z) g^* \circ \varphi(z) \tag{2*}$$

on the one hand. Since $\mathfrak{R}$ allows a strong lumpation and since $g^* \circ \varphi \in B(E, \mathbb{C})$, on the other hand we have

$$\sum_{z \in E} p(y,d,z)(g^* \circ \varphi)(z) = \sum_{z \in E} p(x,d,z)(g^* \circ \varphi)(z) \geq$$

$$\geq \sum_{z \in E} p(x,d',z)(g^* \circ \varphi)(z) =$$

$$= \sum_{z \in E} p(y,d',z)(g^* \circ \varphi(z)),$$

because $\varphi(x) = \varphi(y)$. Obviously, (2*) is an equation and $d \in D^*_{-1}(y)$. Since d is an arbitrary decision and x,y are arbitrary states we obtain the identity $D^*_{-1}(x) = D^*_{-1}(y)$. Consequently,

$$\bigcap_{x \in \varphi^{-1}(s)} D^*_{-1}(x) \neq \emptyset$$

and $\mathbb{D}^{\mathfrak{C}*}_{-1} := \mathbb{D}^{\mathfrak{C}} \cap \mathbb{D}_{-1} \neq \emptyset$. Thus (1*) is proved.

Obviously, $\Theta\mathbb{D}'_{-1} = \mathbb{D}^{\mathfrak{C}*}_{-1}$. By appling the known procedure of proving from (3.16) we obtain

$$g^* \circ \varphi + v^* \circ \varphi = \max_{h \in \mathbb{D}^{\mathfrak{C}*}_{-1}} \{r(h) + P(h)(v^* \circ \varphi)\}.$$

In the following we will show that

$$\max_{h \in \mathbb{D}^{\mathfrak{C}*}_{-1}} \{r(h) + P(h)(v^* \circ \varphi)\} = \max_{h \in \mathbb{D}^*_{-1}} \{r(h) + P(h)(v^* \circ \varphi)\}. \qquad (3^*)$$

From this (3.18) will follow. Let

$$D^*_0(x) := \{d \in D^*_{-1}(x) : r(x,d) + \sum_{z \in E} p(x,d,z)v^* \circ \varphi(z) =$$

$$= \max_{d' \in D^*_{-1}(x)} \{r(x,d') + \sum_{z \in E} p(x,d',z)v^* \circ \varphi(z)\}\},$$

$x \in E$. Further, let x,y be any states with $\varphi(x) = \varphi(y)$ and let $d \in D^*_0(x)$, $d' \in D^*_0(y)$. Based on the results proved above we have

$$d, d' \in D^*_{-1}(x) = D^*_{-1}(y).$$

Moreover, due to the definition

$$r(y,d) + \sum_{z \in E} p(y,d,z)v^* \circ \varphi(z) \leqq r(y,d') + \sum_{z \in E} p(y,d',z)v^* \circ \varphi(z). \quad (4^*)$$

Since $v^* \circ \varphi \in B(E, \mathfrak{C})$ is real and $\mathfrak{M}$ is strongly lumpable we obtain the inequality

$$r(y,d) + \sum_{z \in E} p(y,d,z)v^* \circ \varphi(z) = r(x,d) + \sum_{z \in E} p(x,d,z)v^* \circ \varphi(z) \geqq$$

$$\geqq r(x,d') + \sum_{z \in E} p(x,d',z)v^* \circ \varphi(z) = r(y,d') + \sum_{z \in E} p(y,d',z)v^* \circ \varphi(z).$$

This implies that (4*) must be an equation. I.e., it holds that $d \in D^*(y)$. Hence $D^*_0(x) = D^*_0(y)$, because x,y are arbitrary states. Consequently,

$$\bigcap_{x \in \varphi^{-1}(s)} D^*_0(x) \neq \emptyset,$$

i.e., there is a function $f \in \mathbb{D}^{\mathfrak{C}*}_{-1}$ attaining $\max_{h \in \mathbb{D}^*_{-1}} \{r(h) + P(h)(v^* \circ \varphi)\}$.

Due to this consideration (3*) is proved.
Finally, let f* be defined as in (i). Then it is easy to see that the relations

$$g^* \circ \varphi = P(\theta f)(g^* \circ \varphi) \text{ and}$$

$$g^* \circ \varphi + v^* \circ \varphi = r(\theta f^*) = P(\theta f^*)(v^* \circ \varphi)$$

are verified. In view of theorem 2.26 $(\theta f^*)^\infty = \hat{\theta}\tau^*$ is an average optimal policy on $\mathfrak{R}$. ∎

Up to now we have not been able to prove the optimality of $\hat{\theta}\tau^*$ on $\mathfrak{R}$ without the finiteness assumption with respect to E, though the system (3.15) - (3.16) has a solution. We shall show in the sequel that, nevertheless, $\hat{\theta}\tau^*$ is optimal on $\mathfrak{R}_s$ (instead on $\mathfrak{R}_s^{\mathfrak{C}}$) if the supposition (v) is fulfilled. This result sharpens the theorem 3.11, too.
If we consider the proofs of the last two theorems we notice that the validity of the so-called BELLMAN equations (3.14), (3.15), and (3.16), respectively, plays an important part regarding the optimality of the $\mathfrak{C}$-compatible policies in $\mathfrak{R}$ (instead of $\mathfrak{R}^{\mathfrak{C}}$) which correspond to the optimal policies in $\mathfrak{M}'$ (see also 3.4).

Some authors found sufficient conditions for the solvability of the system (3.14), which were already mentioned in the section 2.4. The theorem 3.19 is valid under these conditions, i.e., the exclusive consideration of piecewise constant policies of the original model does not mean any restriction. However, as already mentioned in remark 2.14, there are results concerning the existence of an average optimal policy without regard to any BELLMAN equation (for example, see HORDIJK [54], [55] and WIJNGAARD [121] - [124]). The authors do not know whether the optimality on $\mathfrak{R}^{\mathfrak{C}}$ (and $\mathfrak{R}_s^{\mathfrak{C}}$, respectively) coincides with the optimality on $\mathfrak{R}$ (and $\mathfrak{R}_s$, respectively) in these cases. The solution of this problem seems to be difficult.

THE HOWARD ALGORITHM WITH RESPECT TO THE AVERAGE REWARD

Now we consider how the policy improvement procedure given in 2.4 operates in the model $\mathfrak{M}$ as well as in the model $\mathfrak{M}'$. According to theorem 2.17 the HOWARD-algorithm can be applied within the class $\mathfrak{R}_s'$ if the following assumptions are fulfilled:
($\bar{V}$) $\bar{P}(f)$ is 1-recurrent for every $f \in \mathbb{D}'$ and the $(P,r)^*$-equations

$$\begin{aligned} g &= \bar{P}(f)g, \\ g + v &= \bar{r}(f) + \bar{P}(f)v, \\ \bar{P}(f)^* v &= 0 \end{aligned} \qquad (3.19)$$

have a solution in B(E').
(If there is a solution of (3.19), then it is uniquely determined and will be denoted by (g^f, v^f), cf. section 2.3.)

For any pair of functions (f,g) with $f,g \in \mathbb{D}'$ we can introduce the following terms

$$\bar{\phi}^{-1}(g,f) := (\bar{P}(g) - I)\bar{g}^{f},$$

$$\bar{\phi}^{o}(g,f) := \bar{r}(g) + (\bar{P}(g) - I)\bar{v}^{f} - \bar{g}^{f}, \text{ and}$$

$$\bar{\Psi}^{o}(g,f) := (\bar{\phi}^{-1}(g,f), \bar{\phi}^{o}(g,f)).$$

It is easy to see that $(\bar{V})$ does not imply the validity of (V) for the original model (see section 2.4). However, the following lemma holds.

<u>Lemma 3.21.</u> The condition

$(V_{\mathbb{C}})$: $P(h)$ is 1-recurrent for each $h \in \mathbb{D}^{\mathbb{C}}$, and the $(P,r)^*$-equations are solvable with respect to h

is sufficient for $(\bar{V})$ to be fulfilled. In particular,

$$g^{f} \circ \varphi = g^{\theta f} \quad \text{and} \quad v^{f} \circ \varphi = v^{\theta f}$$

for each $f \in \mathbb{D}'$.

<u>Proof:</u> Let $(V_{\mathbb{C}})$ be fulfilled, and let $f \in D'$. Denote $h := \theta f$. Since $P(h)$ is non-dissipative, $P(f)$ is also non-dissipative due to lemma 1.31. Moreover, lemma 1.31 implies that the inclusion $T \supset \varphi^{-1}(T')$ holds for the sets T and T' of transient states (with respect to $P(h)$ and $P(f)$, respectively). One obtains for each state $s \in E'$ being transient with respect to $P(f)$:

$$\sum_{n=0}^{\infty} \bar{P}(f)^{n}_{T'T'} \mathbf{1}_{T'}(s) = \sum_{n=0}^{\infty} \bar{P}(f)^{n}(s,T')$$

$$= \sum_{n=0}^{\infty} P(h)^{n}(x,\varphi^{-1}(T'))$$

$$\leq \sum_{n=0}^{\infty} P(h)^{n}(x,T) = \sum_{n=0}^{\infty} P(h)^{n}_{TT}\mathbf{1}(x) < \infty$$

for every $x \in \varphi^{-1}(s)$. Thus $\bar{P}(f)$ is $\mathbf{1}$-recurrent. Now it remains to prove that the $(P,r)^*$-equations possess a solution with respect to f. As supposed, these equations are solvable with respect to h and have the solutions g^{h}, v^{h}. Following theorem 3.9 $g^{h} \in B(E, \mathfrak{C})$. The function v^{h} also belongs to $B(E, \mathfrak{C})$ as the following considerations will show: Due to (2.7) and (2.8) we have the equations

$$g^{h} + P(h)^{k}v^{h} = P(h)^{k}g^{h} + P(h)^{k}v^{h} = P(h)^{k}r(h) + P(h)^{k+1}v^{h}$$

for each $k \in \mathbb{N}_{o}$, and, consequently,

$$ng^{h} - \sum_{k=0}^{n-1} P(h)^{k}r(h) = P(h)^{n}v^{h} - v^{h}.$$

We obtain by summarizing these terms with respect to n and by dividing by N:

$$v_N := \frac{1}{N}\sum_{n=1}^{N}(ng^h - \sum_{k=0}^{n-1} P(h)^k r(h)) = \frac{1}{N}\sum_{n=1}^{N} P(h)^n v^h - v^h .$$

Obviously, each of the functions v_N, $N \in \mathbb{N}$, belongs to $B(E, \mathfrak{C})$. According to the assumption P(h) is non-dissipative. Taking lemma 1.4 and (2.9) into account we see that

$$\lim_{N\to\infty} v_N(x) = \lim_{N\to\infty}\frac{1}{N}\sum_{n=1}^{N} P(h)^n v^h(x) - v^h(x)$$

$$= P(h)^* v^h(x) - v^h(x) = - v^h(x).$$

Thus v^h is the componentwise limit on $\mathfrak{C}$-measurable functions and, consequently, it is also $\mathfrak{C}$-measurable.

The functions $\bar{g}^f = \Phi^{-1}g^h$ (cf. (3.12)) and $\bar{v} := \Phi^{-1}v^h$ form a solution of the $(P,r)^*$-equations in $B(E')$ with respect to f as the following calculation will show:

Namely, due to (2.6) $\bar{g}^f = \bar{P}(f)^*\bar{r}(f)$ and, consequently,

$$\bar{P}(f)\bar{g}^f = \bar{P}(f)\bar{P}(f)^*\bar{r}(f) = \bar{P}(f)^*\bar{r}(f) = \bar{g}^f$$

by using lemma 1.4. Further, (3.5), (3.6), (3.12), and the definition of $\tilde{\Phi}$ imply

$$(\bar{g}^f + \bar{v})\circ\varphi = g^h + v^h = r(h) + P(h)v^h =$$

$$= \Phi(\bar{r}(f)) + \tilde{\Phi}(\bar{P}(f))(\Phi\bar{v}) = \Phi(\bar{r}(f) + \bar{P}(f)\bar{v}) =$$

$$= (\bar{r}(f) + \bar{P}(f)\bar{v})\circ\varphi.$$

Since φ is surjective the equation

$$g^f + v^f = \bar{r}(f) + \bar{P}(f)v$$

follows. Using lemma 1.3 we obtain the following equation in a similar way:

$$(\bar{P}(f)^*\bar{v})\circ\varphi = \tilde{\Phi}(\bar{P}(f)^*)(\Phi\bar{v}) = P(h)^* v^h = 0.$$

This implies that $\bar{P}(f)^*\bar{v} = 0$ due to the surjectivity of φ. It follows that the $(P,r)^*$-equations are solvable with respect fo f. In particular,

$$\bar{g}^f\circ\varphi = g^{\theta f} \text{ and } v^f\circ\varphi = v^{\theta f}. \blacksquare$$

<u>Note:</u> If E is finite, then the assumptions $(\bar{V})$ and (v) are automatically fulfilled and hence $(V_{\mathfrak{C}})$ is true.

With the notations

$$\bar{\mathfrak{G}}_o(f) := \{g \in \mathbb{D}' : \bar{\Psi}^o(g,f) \succ 0\} \text{ and}$$

$$\bar{\Psi}^o(g,f)\circ\varphi := (\bar{\psi}^{-1}(g,f)\circ\varphi,\ \bar{\psi}^o(g,f)\circ\varphi)$$

$(g,f \in \mathbb{D}')$ we obtain

Lemma 3.22. If $(V_{\mathfrak{C}})$ is fulfilled, then

(i) $\bar{\psi}^i(g,f) \circ \varphi = \psi^i(\theta g, \theta f)$ $(i = -1, 0)$ and

(ii) $\bar{\Psi}^o(g,f) \circ \varphi = \Psi^o(\theta g, \theta f)$

for $f,g \in \mathbb{D}'$.

Furthermore, for any state $s \in E'$ and all $x \in \varphi^{-1}(s)$,

(iii) $\bar{\psi}^i(g,f)(s) \geq 0$ (> 0 resp.) $\Longleftrightarrow$ $\psi^i(\theta g, \theta f)(x) \geq 0$ (> 0 resp.), $i = -1, 0$, and

(iv) $\bar{\Psi}^o(g,f)(s) \succcurlyeq 0$ ($\succ 0$ resp.) $\Longleftrightarrow \Psi^o(\theta g, \theta f)(x) \succcurlyeq 0$ ($\succ 0$ resp.).

(v) $g \in \bar{\mathfrak{G}}_o(f) \Longleftrightarrow \theta g \in \mathfrak{G}_o(\theta f)$.

Proof: The lemma 3.21 and (3.6) imply

$$\bar{\psi}^{-1}(g,f) \circ \varphi = ((\bar{P}(g) - I)\bar{g}^f) \circ \varphi = (\tilde{\Phi}\bar{P}(g) - I)(\bar{g}^f \circ \varphi) =$$
$$= (P(\theta g) - I)g^{\theta f} = \psi^{-1}(\theta g, \theta f).$$

The assertion for $i = 0$ can analogously be proved, where (3.5) is applied. The propositions (ii), (iii), (iv), and (v) are evident consequences.

We consider the following statements in order to prepare the next theorems, whereas the assumption has to be fulfilled for any state $s \in E'$:

$\bar{A}(s)$: "There exists a state s' being recurrent (with respect to g) with $s \overset{g}{\curvearrowright} s'$ and $\bar{\psi}^o(g,f) > 0$".

$\bar{B}(s)$: "There exists a state $s' \in E'$ with $s \overset{g}{\curvearrowright} s'$ and $\bar{\psi}^{-1}(g,f)(s') > 0$".

$\bar{C}(s)$: "For every state $s' \in E'$ with $s \overset{g}{\curvearrowright} s'$ the equation $\bar{\psi}^{-1}(g,f)(s') = 0$ holds, and $\bar{\psi}^o(g,f)(s'') = 0$ for each recurrent state s'' (with respect to g) with $s \overset{g}{\curvearrowright} s''$".

$\bar{C}^*(s)$: "$\bar{C}(s)$ holds, and there exists a state s' with $s \overset{g}{\curvearrowright} s'$ and $\bar{\psi}^o(g,f)(s') > 0$".

Furthermore, let $A(x)$, $B(x)$, $C(x)$, and $C^*(x)$ be statements analogously defined as $\bar{A}(s)$, $\bar{B}(s)$, $\bar{C}(s)$, and $\bar{C}^*(s)$, respectively, where s, g, $\bar{\psi}^{-1}$ and $\bar{\psi}^o$ are replaced by x, θg, ψ^{-1}, and ψ^o, respectively (cf. section 2.4).

On the basis of the propositions of the preceding lemmas we obtain

Theorem 3.23. Let $f,g \in \mathbb{D}'$. If $(V_{\mathfrak{C}})$ is fulfilled, then the following relations are true for any state $s \in E'$ and all $x \in \varphi^{-1}(s)$.

(i) $\bar{A}(s) \Longleftrightarrow A(x)$,
$\bar{B}(s) \Longleftrightarrow B(x)$,
$\bar{C}(s) \Longleftrightarrow C(x)$, and
$\bar{C}^*(s) \Longleftrightarrow C^*(x)$.

(ii) $\bar{\Psi}^0(g,f) \geq 0 \Rightarrow \bar{g}^g \geq \bar{g}^f$

$$\Updownarrow \qquad\qquad \Updownarrow$$

$\Psi^0(\theta g,\theta f) \geq 0 \Rightarrow g^{\theta g} \geq g^{\theta f}$.

If $\Psi^0(g,f) \geq 0$, then

(iii) $\bar{A}(s) \vee \bar{B}(s) \Longleftrightarrow \bar{g}^g(s) > \bar{g}^f(s)$

$$\Updownarrow \qquad\qquad \Updownarrow$$

$A(s) \vee B(x) \Longleftrightarrow g^{\theta g}(x) > g^{\theta f}(x)$, and

(iv) $\bar{C}(s) \Longleftrightarrow \bar{g}^g(s) = \bar{g}^f(s)$

$$\Updownarrow \qquad\qquad \Updownarrow$$

$C(x) \Longleftrightarrow g^{\theta g}(x) = g^{\theta f}(x)$.

If g satisfies the additional condition

$$\bar{\Psi}^0(g,f)(s) = 0 \Rightarrow g(s) = f(s) \qquad (*)$$

for all $s \in E'$, then

(v) $\bar{C}(s) \Longleftrightarrow \bar{g}^g(s) = \bar{g}^f(s) \wedge \bar{v}^g(s) \geq \bar{v}^f(s)$

$$\Updownarrow \qquad\qquad \Updownarrow$$

$C(x) \Longleftrightarrow g^{\theta g}(x) = g^{\theta f}(x) \wedge v^{\theta g}(x) \geq v^{\theta f}(x)$

and

(vi) $\bar{C}^*(s) \Longleftrightarrow g^g(s) = g^f(s) \wedge v^g(s) > v^f(s)$ and s is transient with respect to g

$$\Updownarrow \qquad\qquad \Updownarrow$$

$C^*(x) \Longleftrightarrow g^{\theta g}(x) = g^{\theta f}(x) \wedge v^{\theta g}(x) > v^{\theta f}(x)$ and x is transient with respect to θg.

<u>Proof:</u> Firstly, let us remark that the condition $(V_{\mathfrak{C}})$ and the proposition of lemma 3.21 ensures that the assertions $\bar{A}(s),\ldots,\bar{C}^*(s)$ and $A(x),\ldots,C^*(x)$ make sense, i.e., all terms are well-defined.

(i) "$A(x) \Longleftrightarrow \bar{A}(s)$": Let $A(x)$ hold, i.e., there exists a recurrent state x' (with respect to θg) with $x \overset{\theta g}{\curvearrowright} x'$ and $\psi^0(\theta g,\theta f)(x') > 0$. Let $s' := \varphi(x)$. Due to the lemma 1.30 the relation $s \overset{g}{\curvearrowright} s'$ holds. Lemma 1.31 implies that s' is a recurrent state with respect to g. Furthermore, due to lemma 3.22

$$\bar{\psi}^0(g,f)(s') = \psi^0(\theta g,\theta f)(x) > 0.$$

Thus $\bar{A}(s)$ is true.

On the other hand, if $\bar{A}(s)$ is true, then there exists a recurrent state s' (with respect to g) with $s \overset{g}{\curvearrowright} s'$ and $\bar{\psi}^0(g,f)(s') > 0$. Consequently, there are a state $x' \in \varphi^{-1}(s)$ with $x \overset{\theta g}{\curvearrowright} x'$ (due to the lemma 1.30) and a state $x'' \in \varphi^{-1}(s')$ being recurrent with respect to θg (due to the lemma 1.31) with $x' \overset{\theta g}{\curvearrowright} x''$. Summarizing these properties we obtain: $x \overset{\theta g}{\curvearrowright} x''$, x'' is recurrent with respect to θg and $\psi^o(\theta g,\theta f)(x'') = \bar{\psi}^o(g,f)(s') > 0$ (due to the lemma 3.22). I.e., $A(x)$ holds.

"$B(x) \Longleftrightarrow \bar{B}(s)$" can be proved in an analogous way. By the following equivalence relations

$$C(x) \Longleftrightarrow \neg\, (A(x) \vee B(x))$$

and

$$\bar{C}(s) \Longleftrightarrow \neg\, (\bar{A}(s) \vee \bar{B}(s)),$$

respectively, we immediately obtain $C(x) \Longleftrightarrow \bar{C}(s)$.

"$C^*(x) \Longleftrightarrow \bar{C}^*(s)$": Let $C^*(x)$ be true. Then $C(x)$ holds. Additionally, there is a state x' with $x \overset{\theta g}{\curvearrowright} x'$ and $\psi^o(\theta g,\theta f)(x') > 0$. Then $\bar{C}(s)$ also holds as we have just shown. Furthermore, we obtain the relations $s \overset{g}{\curvearrowright} s' := \varphi(x')$ (due to the lemma 1.30) and $\bar{\psi}^o(g,f)(s') = \varphi^o(\theta g,\theta f)(x') > 0$ (due to the lemma 3.22).
Reversely, if $\bar{C}^*(s)$ holds, then $\bar{C}(s)$ and, consequently, $C(x)$ hold. Furthermore, there is a state s' with $s \overset{g}{\curvearrowright} s'$ and $\psi^o(g,f)(s') > 0$. Then there is a state $x' \in \varphi^{-1}(s')$ with $x \overset{\theta g}{\curvearrowright} x'$ (due to the lemma 1.30). The lemma 3.22 implies that $\psi^o(\theta g,\theta f)(x') = \psi^o(g,f)(s') > 0$. Thus $C^*(x)$ is true and completes the proof of (i).

In order to simplify the proof of (ii) - (vi) we introduce the following enumeration of arrows in the corresponding diagrams:

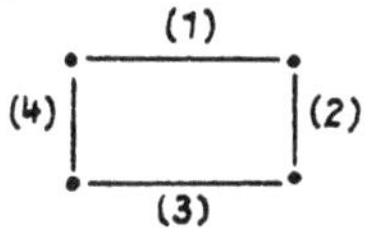

The equivalence relation (4) in (ii) and (iii) - (vi), respectively, immediately follows from lemma 3.22 and from (i), respectively. The implications (1) and (3) can be obtained by a direct application of the related assertions of theorem 2.17 to the transformed and the original model, respectively, in all diagrams. Thereby, it should be noted that the condition (*) is equivalent to the following one:

$$\Psi^o(\theta g, \theta f)(x) = 0 \Longrightarrow \theta g(x) = \theta f(x), \; x \in E.$$

Thus the equivalence relations (2) with respect to (iii) - (vi) are obviously true.
Finally, the relation (2) with respect to (ii) follows directly from theorem 3.9. ∎

Theorem 3.23 illustrates the very close connection between the models $\mathfrak{M}$ and $\mathfrak{M}'$ with respect to the application of the theorem 2.17. As already remarked (see section 2.4) theorem 2.17 founds the policy improvement procedure. Hence, theorem 3.23 shows that we can expect a proper policy improvement step within the transformed model, if and only if we can expect a proper increase of the average reward in each of the states

$x \in \varphi^{-1}(s)$ using the corresponding policy improvement step within the original model. If "good decisions" are maintained (i.e., the condition (*) is fulfilled), then the same can be said of the parallel improvement of the corresponding "weights" v^g and $v^{\theta g}$, respectively.

The corollaries described in section 2.4 and which result from the theorem 2.17 could also be discussed in detail with respect to the original model and the transformed one. However, we restrict ourselves to an assertion concerning a possible break-off at the improvement algorithm.

<u>Lemma 3.24.</u> Let $h_o \in \mathbb{D}^{\mathfrak{C}}$ be a decision function whose (P,r^*)-equations are solvable. Then

$$\max_{h\in\mathbb{D}} \Psi^o(h,h_o) = \max_{h\in\mathbb{D}^{\mathfrak{C}}} \Psi^o(h,h_o). \quad {}^{1)}$$

<u>Proof:</u> Let $D_{-1}(x) := D_{-1}^{h_o}(x)$ and $D_o(x) := D_o^{h_o}(x)$ for each $x \in E$ (cf. remark 2.12). Let the same be true for $\mathbb{D}_{-1} := \mathbb{D}_{-1}^{h_o}$ and $\mathbb{D}_o := \mathbb{D}_o^{h_o}$. If $\mathbb{D}_{-1}^{\mathfrak{C}} := \mathbb{D}_{-1} \cap \mathbb{D}^{\mathfrak{C}}$ and $\mathbb{D}_o^{\mathfrak{C}} := \mathbb{D}_o \cap \mathbb{D}^{\mathfrak{C}}$, then we only have to show that the set $\mathbb{D}_o^{\mathfrak{C}}$ is non-empty. This will be done in the following.

Let $x,y \in E$ with $\varphi(x) = \varphi(y)$ and $d \in D_{-1}(x)$ be any states and decisions, respectively. Since $\mathfrak{M}$ is strongly lumpable and since $g^{h_o} \in B(E,\mathfrak{C})$ (due to lemma 3.21) we obtain the following relations for $d' \in D_{-1}(y)$:

$$\begin{aligned}
\psi^{-1}(h_o)(y,d) &= \sum_{z\in E} (p(y,d,z) - \delta_{yz})g^{h_o}(z) = \\
&= \sum_{z\in E} (p(x,d,z) - \delta_{xz})g^{h_o}(z) = \psi^{-1}(h_o)(x,d) \geq \\
&\geq \psi^{-1}(h_o)(x,d') = \sum_{z\in E}(p(x,d',z) - \delta_{xz})g^{h_o}(z) = \\
&= \sum_{z\in E}(p(y,d',z) - \delta_{yz})g^{h_o}(z) = \psi^{-1}(h_o)(y,d').
\end{aligned}$$

Consequently, $d \in D_{-1}(y)$. Since x and y are arbitrary elements of E

$$D_{-1}(x) = D_{-1}(y)$$

by a symmetry argument. This implies

$$\bigcap_{x\in\varphi^{-1}(s)} D_{-1}(x) \neq \emptyset$$

and

$$\mathbb{D}_{-1}^{\mathfrak{C}} \neq \emptyset.$$

---

1) "max" is to be understood in the lexicographical sense.

Now we again choose any $x, y \in E$ with $\varphi(x) = \varphi(y)$ and $d \in D_0(x)$. As we have just proved each $d' \in D_0(y)$ also belongs to $D_{-1}(y) = D_{-1}(x)$. Due to lemma 3.21 ($v^{h_0} \in B(E, \mathfrak{C})$) and the strong lumpability of $\mathfrak{M}$ we obtain:

$$\begin{aligned}
\psi^0(h_0)(y,d) &= r(y,d) + \sum_{z \in E} (p(y,d,z) - \delta_{yz})v^{h_0}(z) - g^{h_0}(y) = \\
&= r(x,d) + \sum_{z \in E} (p(x,d,z) - \delta_{xz})v^{h_0}(z) - g^{h_0}(x) = \\
&= \psi^0(h_0)(x,d) \geq \\
&\geq \psi^0(h_0)(x,d') = \\
&= r(x,d') + \sum_{z \in E} (p(x,d',z) - \delta_{xz})v^{h_0}(z) - g^{h_0}(x) = \\
&= r(y,d') + \sum_{z \in E} (p(y,d',z) - \delta_{yz})v^{h_0}(z) - g^{h_0}(y) = \\
&= \psi^0(h_0)(y,d').
\end{aligned}$$

Therefore, $d \in D_0(y)$. Thus we can again conclude that $\mathbb{D}_0^{\mathfrak{C}} \neq \emptyset$. ∎

Note that we did not exclude up to now that the $(P,r)^*$-equations are insoluble or that the condition of 1-recurrenceis not fulfilled with respect to certain policies (of the original model) which are not piecewise constant. If we exclude such a possibility, then a break-offof the algorithm with respect to $\mathfrak{M}'$ provides an average optimal policy τ^* on $\mathfrak{R}_S$ as well as, by transformation of τ^* to $\theta\tau^*$, a policy which is average optimal not only on $\mathfrak{R}_S^{\mathfrak{C}}$ but on $\mathfrak{R}_S$.

<u>Theorem 3.25</u>

(i) If $(V_{\mathfrak{C}})$ is fulfilled, then the following relations are valid for each $f \in \mathbb{D}'$:

$$\begin{array}{ccc}
\overline{\mathfrak{B}}_0(f) = \emptyset & \Longrightarrow & f^\infty \text{ is average optimal on } \mathfrak{R}_S' \\
\Updownarrow & & \Updownarrow \\
\mathfrak{B}_0(\theta f) = \emptyset & \Longrightarrow & (\theta f)^\infty \text{ is average optimal on } \mathfrak{R}_S^{\mathfrak{C}} .
\end{array}$$

(ii) If (v) is fulfilled and $\overline{\mathfrak{B}}_0(f) = \emptyset$ for any $f \in \mathbb{D}'$, then $(\theta f)^\infty$ is an average optimal policy on $\mathfrak{R}_S$.

<u>Proof:</u> Let us enumerate the diagram given in (i) in the following way:

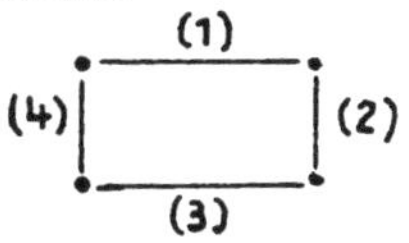

(1) results from lemma 3.21 and theorem 2.20 (i).
The relation (2) is a special case of the assertions of theorem 3.11, (i) and (iii). (3) follows from theorem 2.20 (i). Relation (4) can be proved as follows.
Due to lemma 3.22 and lemma 3.24 we obtain:

$$\overline{\mathfrak{G}}_o(f) = \emptyset \iff \mathfrak{G}_o(\theta f) \cap \mathbb{D}^{\mathfrak{C}} = \emptyset \iff \max_{h \in \mathbb{D}^{\mathfrak{C}}} \Psi^o(h,\theta) = 0$$

$$\iff \max_{h \in \mathbb{D}} \Psi^o(h,\theta f) = 0 \iff \mathfrak{G}_o(\theta f) = \emptyset.$$

Thus (i) is proved.
About (ii): (i) implies that $\mathfrak{G}_o(\theta f) = \emptyset$. Since (v) is supposed to be fulfilled theorem 2.20 (ii) can be applied. ∎

Note: The difference between the assertions (i) and (ii) of the preceding theorem is mainly based on the technique of proving. Namely, if only $(V_{\mathfrak{C}})$ is fulfilled, but (v) is not, then, by theorem 2.20(i), $g^h \leq g^{\theta f}$ (‡) can be concluded from $\psi^o(h, \theta f) \preccurlyeq 0$, if h is piecewise constant. It is unknown yet whether there is the possibility of a certain function $h \in \mathbb{D} \setminus \mathbb{D}^{\mathfrak{C}}$ fulfilling (*) but not (‡).

3.4. SENSITIVE CRITERIA OPTIMIZATION

The assertions concerning the sensitive discount criteria are fully analogous to those concerning the average reward criterion. As described in section 2.6 existence considerations play a subordinated role here. Hence we restrict ourselves to examining how the HOWARD-type policy improvement procedure operates with respect to $\mathfrak{M}$ and $\mathfrak{M}'$. We represent the parallel optimality assertions within the class of stationary and of Markov policies, respectively.

For this purpose we refer to the Laurent expansion of the total reward introduced in section 2.5. Firstly, we consider the relations between the coefficients of these expansions with respect to the two models $\mathfrak{M}$ and $\mathfrak{M}'$.

Let $f \in \mathbb{D}'$ and let $P(\theta f)$ be quasi-compact. Then, due to lemma 1.28, P(f) is also quasi-compact. There exist

$$\overline{P}(f)^* := \lim_{N\to\infty} \frac{1}{N} \sum_{n=0}^{N-1} \overline{P}(f)^n,$$

$$\overline{Z} := (I - (\overline{P}(f) - \overline{P}(f)^*))^{-1}, \text{ and}$$

$$\overline{H}(f) := \overline{Z}(f) - \overline{P}(f)^*$$

both as linear operators on $B(E')$ and as $E' \times E'$-matrices (cf. 1.3 and 1.4). Due to theorem 2.27 we obtain

$$\bar{V}_\beta^f = (1+\rho) \sum_{n=-1}^{\infty} \rho^n \bar{v}_n^f \qquad (3.20)$$

with

$$\bar{v}_{-1}^f := \bar{P}(f)^* \bar{r}(f) \text{ and } \bar{v}_n^f := (-1)^n \bar{H}(f)^{n+1}\, \bar{r}(f),$$

$n \in \mathbb{N}_0$, for every $\beta \in (0,1]$ and $\rho := \beta^{-1}(1-\beta)$, where $V_\beta^{\theta f}$ has the Laurent expansion

$$V_\beta^{\theta f} = (1+\rho) \sum_{n=-1}^{\infty} \rho^n\, v_n^{\theta f}. \qquad (3.21)$$

Lemma 3.26. Let $f \in \mathbb{D}'$ and let $P(\theta f)$ be quasi-compact. If $\beta \in (0,1)$ then the coefficients of the expansions (3.20) and (3.21) fulfil the equations:

$$v_n^{\theta f} = \bar{v}_n^f \circ \varphi = \Phi\, \bar{v}_n^f \quad (n \in \mathbb{N}_{-1}). \qquad (3.22)$$

Proof: Due to theorem 1.29

$$\tilde{\Phi}\bar{P}(f)^* = P(\theta f)^*_{\mathbb{C}} \text{ and } \tilde{\Phi}\bar{H}(f) = H(\theta f)_{\mathbb{C}}$$

since $r(\theta f) \in B(E, \mathbb{C})$. Taking (3.5) into account we obtain

$$\Phi\, \bar{v}_{-1}^f = \Phi(\bar{P}(f)^* \bar{r}(f)) = (\tilde{\Phi}\bar{P}(f)^*)(\Phi\bar{r}(f)) = P(\theta f)^*_{\mathbb{C}} r(\theta f) =$$

$$= P(\theta f)^* r(\theta f) = v_{-1}^{\theta f},$$

$n \in \mathbb{N}_0$, and

$$\Phi\, \bar{v}_n^f = \Phi((-1)^n\, \bar{H}(f)^{n+1}\bar{r}(f)) = (-1)^n(\tilde{\Phi}\bar{H}(f))^{n+1}(\Phi\bar{r}(f)) =$$

$$= (-1)^n\, H(\theta f)_{\mathbb{C}}^{n+1} r(\theta f) = (-1)^n\, H(\theta f)^{n+1} r(\theta f) = v_n^{\theta f}. \blacksquare$$

Let $\phi^n(g,h)$, $\Psi^n(g,h)$, and $\mathfrak{b}_n(h)$ as well as $\mathfrak{G}_n(h)$, $g \in \mathbb{D}$, $n \in \mathbb{N}_{-1}$, defined as in 2.5 for every decision function for which the related function V_β^h possesses a Laurent expansion due to theorem 2.27. If $\bar{P}(f)$, $f \in \mathbb{D}'$, is quasi-compact, then the related quantities with respect to the model $\mathfrak{M}'$ will be denoted by $\bar{\phi}^n(g,f)$, $\bar{\Psi}^n(g,f)$, $\bar{\mathfrak{b}}_n(f)$, and $\bar{\mathfrak{G}}_n(f)$, respectively. Furthermore, let

$$\bar{\Psi}^n(g,f) \circ \varphi := (\bar{\phi}^{-1}(g,f) \circ \varphi, \ldots, \bar{\phi}^n(g,f) \circ \varphi) \quad (g \in \mathbb{D}',\ n \in \mathbb{N}_{-1}).$$

Lemma 3.27. Let $f, g \in \mathbb{D}'$ and let $P(\theta f)$ be quasi-compact. Then the following equations hold for $n \in \mathbb{N}_{-1}$:

(i) $\bar{\phi}^n(g,f) \circ \varphi = \phi^n(\theta g, \theta f)$,

(ii) $\bar{\Psi}^n(g,f) \circ \varphi = \Psi^n(\theta g, \theta f)$.

Furthermore, for each state $s \in E'$ and all $x \in \varphi^{-1}(s)$

(iii) $\overline{\Psi}^n(g,f)(s) \geqslant 0$ (>0, resp.) $\Longleftrightarrow \Psi^n(\theta g, \theta f)(x) \geqslant 0$ (>0, resp.).

(iv) $g \in \overline{\mathfrak{G}}_n(f) \Longleftrightarrow \theta g \in \mathfrak{G}_n(\theta f)$.

Proof:

(i) If we define $\overline{r}_n(g) := 0$ ($n \in \mathbb{N}_{-1} \setminus \{0\}$) and $\overline{r}_o(g) := \overline{r}(g)$ and $r_n(\theta g)$ in an analogous manner, then (3.5), (3.6) and (3.22) imply

$$\overline{\Psi}^n(g,f) \circ \varphi = (\overline{r}_n(g) + (\overline{P}(g) - I)\overline{v}_n^f - \overline{v}_{n-1}^f) \circ \varphi =$$

$$= \overline{r}_n(g) \circ \varphi + \tilde{\Phi}(\overline{P}(g) - I)(\overline{v}_n^f \circ \varphi) - \overline{v}_{n-1}^f \circ \varphi =$$

$$= r_n(\theta g) + (P(\theta g) - I)v_n^{\theta f} - v_{n-1}^{\theta f} = \psi^n(\theta g, \theta f),$$

where $v_{-2}^f := 0$ and $v_{-2}^{\theta f} := 0$.

The remaining assertions are immediate consequences of (i). ∎

Now we shall prove a result which is analogous to that of theorem 3.23. For this purpose let $P(\theta f)$ and $P(\theta g)$, $f,g \in \mathbb{D}'$, be quasi-compact operators. Firstly, we introduce some useful notations. Let $\overline{R}$ be the set of recurrent states in E' with respect to g and $\overline{T} := E' \setminus \overline{R}$. Further we denote for each $s \in E'$:

$$\overline{R}(s) := \{s' \in \overline{R} : s \overset{g}{\frown} s'\},$$

$$\overline{T}(s) := \{s' \in \overline{T} : s \overset{g}{\frown} s'\}, \text{ and}$$

$$\overline{E}(s) := \overline{R}(s) \cup \overline{T}(s).$$

$N \in \mathbb{N}_{-1}$ is fixed in the following.

Let us consider the following statements:

$$\overline{C}_k(s) :\Longleftrightarrow \overline{\psi}^{-2}(g,f)_{\overline{E}(s)} = \dots = \overline{\psi}^{k-1}(g,f)_{\overline{E}(s)} = 0 \text{ and }^{1)}$$
$$\overline{\psi}^k(g,f)_{\overline{R}(s)} = 0,$$

$$\overline{A}_k(s) :\Longleftrightarrow \overline{C}_k(s) \wedge (\overline{\psi}^{k+1}(g,f)_{\overline{R}(s)} > 0),$$

$$\overline{B}_k(s) :\Longleftrightarrow \overline{C}_k(s) \wedge (\overline{\psi}^{k}(g,f)_{\overline{T}(s)} > 0)$$

and [2])

$$\overline{A}^*(s) :\Longleftrightarrow \bigvee_{k=-1}^{N} \overline{A}_k(s) \text{ as well as } \overline{B}^*(s) :\Longleftrightarrow \bigvee_{k=-1}^{N} \overline{B}_k(s).$$

---

[1]) $\overline{\psi}^{-2}(g,f) := 0$

[2]) If $\overline{T}(s) = \emptyset$, then we define $\overline{B}_k(s) := \emptyset$ (logical zero)

In a quite analogous manner we formulate the assertions $C_k(x)$, $A_k(x)$, $B_k(x)$, $(k = -1, \ldots, N+1)$, and $A^*(x)$, $B^*(x)$, where g, f, $\bar{\phi}^k$, $\bar{R}$, $\bar{T}$, and E' are replaced by θg, θf, ϕ^k $(k \in \mathbb{N}_{-2})$, R, T, and $\mathbb{E}$, respectively (see 2.5).

<u>Theorem 3.28.</u> Let $P(\theta f)$ and $P(\theta g)$, $f,g \in \mathbb{D}'$, be quasicompact. Then the following implications are true:

$$
\text{(i)}\quad \begin{array}{ccc}
\bar{\Psi}^{N+1}(g,f) \succcurlyeq 0 & \Longrightarrow & \bar{\mathfrak{v}}_N(g) \succcurlyeq \bar{\mathfrak{v}}_N(f) \\
\Updownarrow & & \Updownarrow \\
\Psi^{N+1}(\theta g,\theta f) \succcurlyeq 0 & \Longrightarrow & \mathfrak{v}_N(\theta g) \succcurlyeq \mathfrak{v}_N(\theta f).
\end{array}
$$

In the following let the condition $\bar{\Psi}^{N+1}(g,f) \succcurlyeq 0$ be fulfilled. Then the following equivalence relations hold for each state $s \in E'$ and all $x \in \varphi^{-1}(s)$:

$$
\text{(ii)}\quad \begin{array}{ccl}
\bar{C}_k(s) & \Longleftrightarrow & \bar{\mathfrak{v}}_{k-1}(g)(s) = \bar{\mathfrak{v}}_{k-1}(f)(s) \\
& & \wedge\ \bar{\mathfrak{v}}_k(g)(s) \geqslant \bar{\mathfrak{v}}_k(f)(s) \\
\Updownarrow & & \qquad\qquad \Updownarrow \\
C_k(x) & \Longleftrightarrow & \mathfrak{v}_{k-1}(\theta g)(x) = \mathfrak{v}_{k-1}(\theta f)(x) \\
& & \wedge\ \mathfrak{v}_k(\theta g)(x) \geqslant \mathfrak{v}_k(\theta f)(x),
\end{array}
$$

$$
\text{(iii)}\quad \begin{array}{ccl}
\bar{A}_k(s) \vee \bar{B}_k(s) & \Longleftrightarrow & \bar{\mathfrak{v}}_{k-1}(\theta g)(x) = \bar{\mathfrak{v}}_{k-1}(\theta f)(x) \\
& & \wedge\ \mathfrak{v}_k(\theta g)(x) \succ \mathfrak{v}_k(\theta f)(x) \\
\Updownarrow & & \qquad\qquad \Updownarrow \\
A_k(x) \vee B_k(x) & \Longleftrightarrow & \mathfrak{v}_{k-1}(\theta g)(x) = \mathfrak{v}_{k-1}(\theta f)(x) \\
& & \wedge\ \mathfrak{v}_k(\theta g)(x) \succ \mathfrak{v}_k(\theta f)(x)
\end{array}
$$

for $k = -1, \ldots, N$,

$$
\text{(iv)}\quad \begin{array}{ccc}
\bar{C}_{N+1}(s) & \Longleftrightarrow & \bar{\mathfrak{v}}_N(g)(s) = \bar{\mathfrak{v}}_N(f)(s) \\
\Updownarrow & & \Updownarrow \\
C_{N+1}(x) & \Longleftrightarrow & \mathfrak{v}_N(\theta g)(x) = \mathfrak{v}_N(\theta f)(x)
\end{array}
$$

and

$$
\text{(v)}\quad \begin{array}{ccc}
\bar{A}^*(s) \vee \bar{B}^*(s) & \Longleftrightarrow & \bar{\mathfrak{v}}_N(g)(s) \succ \bar{\mathfrak{v}}_N(f)(s) \\
\Updownarrow & & \Updownarrow \\
A^*(x) \vee B^*(x) & \Longleftrightarrow & \mathfrak{v}_N(\theta g)(x) \succ \mathfrak{v}_N(\theta f)(x).
\end{array}
$$

If the decision function g satisfies the additional assumption

$$
\bar{\Psi}^{N+1}(g,f)(s) = 0 \Longrightarrow g(s) = f(s) \qquad (3.23)
$$

for all $s \in E'$, then

$$
\text{(vi)}\qquad
\begin{array}{ccl}
\bar{C}_{N+1}(s) & \Longleftrightarrow & \bar{\mathfrak{d}}_N(g)(s) = \bar{\mathfrak{d}}_N(f)(s) \wedge \\
 & & \bar{\mathfrak{d}}_{N+1}(g)(s) \geqslant \bar{\mathfrak{d}}_{N+1}(f)(s) \\
\Big\Updownarrow & & \qquad\qquad \Big\Updownarrow \\
C_{N+1}(x) & \Longleftrightarrow & \mathfrak{d}_N(\theta g)(x) = \mathfrak{d}_N(\theta f)(x) \\
 & & \mathfrak{d}_{N+1}(\theta g)(x) \geqslant \mathfrak{d}_{N+1}(\theta f)(x)
\end{array}
$$

and

$$
\text{(vii)}\qquad
\begin{array}{ccl}
\bar{B}_{N+1}(s) & \Longleftrightarrow & \bar{\mathfrak{d}}_N(g)(s) = \bar{\mathfrak{d}}_N(f)(s), \\
 & & \bar{\mathfrak{d}}_{N+1}(g)(s) > \bar{\mathfrak{d}}_{N+1}(f)(s), \\
\Big\Updownarrow & & \text{and } s \text{ is transient under } g \\
 & & \qquad\qquad \Big\Updownarrow \\
B_{N+1}(x) & \Longleftrightarrow & \mathfrak{d}_N(\theta g)(x) = \mathfrak{d}_N(\theta f)(x)\ , \\
 & & \mathfrak{d}_{N+1}(\theta g)(x) = \mathfrak{d}_{N+1}(\theta f)(x), \\
 & & \text{and } x \text{ is transient under } \theta g.
\end{array}
$$

Proof: As we have already done the diagram will be enumerated in the following form:

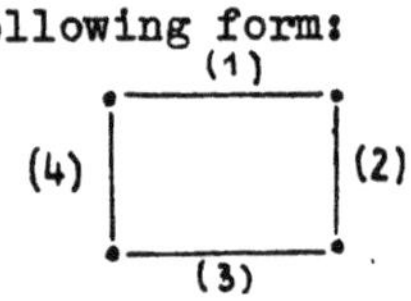

Since not only $P(\theta f)$ and $P(\theta g)$ but also $\bar{P}(f)$ and $\bar{P}(g)$ are quasicompact, we obtain (due to lemma 1.28) the relation (1) and (3) in all diagrams stated in (i) - (vii) as immediate consequences of theorem 2.34. It is to remark that the suppositions (3.23) and (2.27) are obviously equivalent. The relation (4) of (i) follows from lemma 3.27, whereas lemma 3.26 implies (2).
In order to complete the proof we only need to prove that (4) is valid in the remaining diagrams. For this purpose we use three auxiliary assertions:

1°. For each $m \in \mathbb{N}_{-1}$

$$\bar{\psi}^{m}(g,f)_{\bar{E}(s)} = 0 \quad (\geqq 0,\ > 0 \text{ resp.}) \Longleftrightarrow$$

$$\psi^{m}(\theta g,\theta f)_{E(x)} = 0 \quad (\geqq 0,\ > 0 \text{ resp.}).$$

Proof:
"$\Longrightarrow$": Let $\bar{\psi}^{m}(g,f)_{\bar{E}(s)} = 0$ ($\geqq 0$ resp.) and let $x' \in E(x)$ be arbitrary. Further, let $s' := \varphi(x')$. $s \overset{g}{\frown} s'$ follows from lemma 1.30 since $x \overset{g}{\frown} x'$, i.e., $s' \in E(s)$. Therefore, due to lemma 3.27 we obtain the relations:

$$\psi^{m}(\theta g,\theta f)(x') = \bar{\psi}^{m}(g,f)(s') = 0 \ (\geqq 0, \text{ resp.}).$$

" $\Longleftarrow$ ": If $\psi^m(\theta g,\theta f)_{E(x)} = 0$ ($\geqq 0$, resp.) and $s' \in \bar{E}(s)$ (i.e., $s \overset{g}{\curvearrowright} s'$), then, due to lemma 1.30, there is an element $x' \in \varphi^{-1}(s)$ with $x \overset{\theta g}{\curvearrowright} x'$ (i.e., $x' \in E(x)$). Then lemma 3.27 ensures the relations

$$\bar{\psi}^m(g,f)(s') = \psi^m(\theta g,\theta f)(x') = 0 \;(\geqq 0,\text{ resp.}).$$

The logical relation

$$\bar{\psi}^m(g,f)_{\bar{E}(s)} > 0 \Longleftrightarrow \psi^m(\theta g,\theta f)_{E(x)} > 0$$

immediately follows from the two relations we just proved.

2°. For each $m \in \mathbb{N}_{-1}$

$$\bar{\psi}^m(g,f)_{\bar{R}(s)} = 0 \quad (\geqq 0, > 0 \text{ resp.}) \Longleftrightarrow$$

$$\psi^m(\theta g,\theta f)_{R(x)} = 0 \quad (\geqq 0, > 0 \text{ resp.})$$

Proof:

" $\Longrightarrow$ ": If $\psi^m(g,f)_{\bar{R}(s)} = 0$ ($\geqq 0$ resp.) and $x' \in R(x)$ is arbitrarily chosen, then $s \overset{g}{\curvearrowright} s'$ and $s \in \bar{R}(s)$) follow from the relation $x \overset{\theta g}{\curvearrowright} x'$ and from the recurrence of x' with the help of the lemmas 1.30 and 1.31. Moreover, lemma 3.27 implies that

$$\psi^m(\theta g,\theta f)(x') = \bar{\psi}^m(g,f)(s') = 0 \quad (\geqq 0,\text{ resp.}).$$

" $\Longleftarrow$ ": Let $\psi^m(\theta g,\theta f)_{R(x)} = 0$ ($\geqq 0$, resp.) and let s' be an arbitrary state in $\bar{R}(s)$. Then lemma 1.31 guarantees the existence of a state $x' \in \varphi^{-1}(s')$ with $x \overset{\theta g}{\curvearrowright} x'$ and of a state $x'' \in \varphi(s')$ (recurrent with respect to θg) with $x' \overset{\theta g}{\curvearrowright} x''$. Therefore, $x'' \in R(x)$, and lemma 3.27 implies that

$$\bar{\psi}^m(g,f)(s') = \psi^m(\theta g,\theta f)(x'') = 0 \quad (\geqq 0,\text{ resp.}).$$

The assertion with respect to " > " also follows from the preceding considerations.

3°. For each $m \in \mathbb{N}_{-1}$

$$\bar{\psi}^m(g,f)_{\bar{R}(s)} = 0 \wedge \bar{\psi}^m(g,f)_{\bar{T}(s)} > 0 \Longleftrightarrow$$

$$\Longleftrightarrow \psi^m(\theta g,\theta f)_{R(x)} = 0 \wedge \psi^m(\theta g,\theta f)_{T(x)} > 0.$$

Proof:

" $\Longrightarrow$ ": Let $\bar{\psi}^m(g,f)_{\bar{R}(s)} = 0 \wedge \bar{\psi}^m(g,f)_{\bar{T}(s)} > 0$. 2° implies that $\psi^m(\theta g,\theta f)_{R(x)} = 0$. Let $x' \in T(x)$ and $s' := \varphi(x)$. Due to lemma 1.30 $s \overset{g}{\curvearrowright} s'$, and $\psi^m(\theta g,\theta f)(x') = \bar{\psi}^m(g,f)(s')$ by lemma 3.27. Hence, $\psi^m(\theta g,\theta f)(x') = 0$ if s' is recurrent with respect to g. In every case $\psi^m(\theta g,\theta f)(x') \geqq 0$. Furthermore, there exists a transient state s_o with respect to g with $s \overset{g}{\curvearrowright} s_o$ and $\bar{\psi}^m(g,f)(s_o) > 0$. Due to lemma 1.30 we can find a state $x_o \in \varphi^{-1}(s_o)$ with $x \overset{\theta g}{\curvearrowright} x_o$. Lemma 1.31 implies that x_o is transient, whereas $\psi^m(\theta g,\theta f)(x_o) = \bar{\psi}^m(g,f)(s_o) > 0$ follows from lemma 3.27. Altogether we obtain $\psi^m(\theta g,\theta f)_{T(x)} > 0$.

" $\Leftarrow$ ": Reversely, let

$$\psi^m(\theta g,\theta f)_{R(x)} = 0 \wedge \psi^m(\theta g,\theta f)_{T(x)} > 0.$$

$\bar{\psi}^m(g,f)_{\bar{R}(s)} = 0$ because of 2°. Let $s' \in \bar{T}(s)$. Then there is a $x' \in \varphi^{-1}(s')$ with $x \overset{\theta g}{\curvearrowright} x'$. Due to lemma 1.31 x' is transient, and lemma 3.27 shows that

$$\bar{\psi}^m(g,f)(s') = \psi^m(\theta g,\theta f)(x') \geqq 0.$$

On the supposition mentioned above there exists a state $x_o \in T(x)$ with $\psi^m(\theta g,\theta f)(x_o) > 0$. Therefore, $s_o := \varphi(x_o) \in \bar{E}(s)$ and

$$\bar{\psi}^m(g,f)(s_o) = \psi^m(\theta g,\theta f)(x_o) > 0.$$

Since $\bar{\psi}^m(g,f)_{\bar{R}(s)} = 0$ the state s_o must be transient with respect to g. Thus $\bar{\psi}^m(g,f)_{\bar{T}(s)} > 0$, and 3° is proved.

Using the results of 1°, 2°, and 3° we obtain the equivalence relation (4) of (ii) - (v). Firstly, for $m \in \mathbb{N}_{-1}$, we obtain

$$\bar{C}_m(s) \Longleftrightarrow C_m(x). \qquad (4^o)$$

Based on 1° and 2° the following equivalence relations hold:

$$\bar{C}_m(s) \Longleftrightarrow \bar{\psi}^{-2}(g,f)_{\bar{E}(s)} = \dots = \bar{\psi}^{m-1}(g,f)_{\bar{E}(s)} = 0 \wedge$$
$$\wedge \bar{\psi}^m(g,f)_{\bar{R}(s)} = 0$$
$$\Longleftrightarrow \psi^{-2}(\theta g,\theta f)_{E(x)} = \dots = \psi^{m-1}(\theta g,\theta f)_{E(x)} = 0 \wedge$$
$$\wedge \psi^m(\Theta g,\Theta f)_{R(x)} = 0$$
$$\overset{(4^o)}{\Longleftrightarrow} C_m(x).$$

Furthermore, due to (4°), the relations (4) in (ii), (iv), and (vi) are proved.

Furthermore, $\bar{A}_m(s) \Longleftrightarrow A_m(x)$ and $\bar{B}_m(s) \Longleftrightarrow B_m(x)$ for $m \in \mathbb{N}_{-1}$. These relations follow from the following considerations:

$$\bar{A}_m(s) \Longleftrightarrow \bar{C}_m(s) \wedge \bar{\psi}^{m+1}(g,f)_{\bar{R}(s)} > 0$$
$$\Longleftrightarrow C_m(x) \wedge \psi^{m+1}(\theta g,\theta f)_{R(x)} > 0 \quad \text{(due to } 2^o \text{ and } 4^o)$$
$$\Longleftrightarrow A_m(x)$$

and

$$\bar{B}_m(s) \Longleftrightarrow \bar{\Phi}^{-2}(g,f)_{\bar{E}(s)} = \dots = \bar{\Phi}^{m-1}(g,f)_{\bar{E}(s)} = 0 \wedge$$
$$\wedge \bar{\Phi}^m(g,f)_{\bar{R}(s)} = 0 \wedge \bar{\Phi}^m(g,f)_{\bar{T}(s)} > 0$$
$$\Longleftrightarrow \phi^{-2}(\theta g,\theta f)_{E(x)} = \dots = \phi^{m-1}(\theta g,\theta f)_{E(x)} = 0 \wedge$$
$$\wedge \phi^m(\theta g,\theta f)_{R(x)} = 0 \wedge \phi^m(\theta g,\theta f)_{T(x)} > 0$$
(due to 1° and 3°)
$$\Longleftrightarrow B_m(x).$$

Therefore, we have, in particular,

$$\bar{A}_m(s) \vee \bar{B}_m(s) \Longleftrightarrow A_m(x) \vee B_m(x)$$

(i.e., (4) in (iii)). Finally, the equivalence relation

$$\bar{A}^*(s) \vee \bar{B}^*(s) \Longleftrightarrow A^*(x) \vee B^*(x)$$

holds. It proves (4) with respect to (v). The relation

$$\bar{B}_{N+1}(s) \Longleftrightarrow B_{N+1}(x)$$

is valid, i.e. (4) in (vii) is proved. ∎

The theorem just proved admits the following interpretation: Consider the HOWARD algorithm with respect to the transformed model and this algorithmapplied to the class of piecewise constant policies of the original model. If we suppose that the condition ($V^*_{\mathfrak{C}}$): "P(h) is quasi-compact for every $h \in \mathbb{D}^{\mathfrak{C}}$" is fulfilled, then theorem 3.28 implies that there is the same one-to-one correspondence between these algorithms in the case of sensitive criteria as well as of finite stage, discounted and average reward problems. Thus we can expect an essential improvement of a policy within a policy improvement step (i.e., from $f \in \mathbb{D}'$ to $g \in \mathbb{D}'$) with regard to the transformed model in a certain state $s \in E'$, if and only if this happens within the corresponding policy improvement step (i.e., from θf zu θg) with respect to the original for all the states $x \in \varphi^{-1}(s)$. Thus, if we use the HOWARD algorithm, then any policy improvement within the class $\mathfrak{R}^{\mathfrak{C}}_S$ of the original model can be replaced by an improvement of the corresponding decision functions with respect to the transformed model and by a backward transformation of the resulting functions.

Theorem 3.28 enables us to establish an optimality proposition concerning the break-off case of the HOWARD algorithm. For this purpose we need the following

<u>Lemma 3.29.</u> Let $P(\theta f)$, $f \in D'$, be quasi-compact. Then

$$\max_{h\in\mathbb{D}^{\mathfrak{C}}} \Psi^N(h,\theta f) = \max_{h\in\mathbb{D}} \Psi^N(h,\theta f).$$

<u>Proof:</u> Let $f' := \theta f$ and

$$D_{-1}(x) := \{d \in D(x) : \psi^{-1}(\theta f)(x,d) = \max_{d' \in D(x)} \psi^{-1}(\theta f)(x,d')\}$$

and

$$D_n(x) := \{d \in D_{n-1}(x) : \psi^n(\theta f)(x,d) = \max_{d' \in D_{n-1}(x)} \psi^n(\theta f)(x,d')\},$$

for $n \in \mathbb{N}_0$ and $x \in E$,

$$\mathbb{D}_n := \{h \in \mathbb{D} : \psi^n(h,f') = \max_{h' \in \mathbb{D}} \psi^n(h',f')\}, \text{ and}$$

$$\mathbb{D}^{\mathfrak{C}}_n := \mathbb{D}_n \cap \mathbb{D}^{\mathfrak{C}}.$$

Obviously, $\mathbb{D}_n = \prod_{x \in E} D_n(x)$.

In the following we shall prove that $\mathbb{D}^{\mathfrak{C}}_n \neq \emptyset$ for all $N \in \mathbb{N}_{-1}$. (Then, the proposion of the lemma is an immediate consequence).

Firstly, we prove the implication

$$\varphi(x) = \varphi(y) \Longrightarrow D_n(x) = D_n(y) \qquad (*)$$

$(x,y \in E,\ n \in \mathbb{N}_{-1})$.

The following considerations concerning the proof of (*) are based on the strong lumpability of $\mathfrak{M}$ and on the fact that all coefficients $v^{f'}_n$, $n \in \mathbb{N}_{-1}$, of the Laurent expansion of $V^{f'}_\beta$ belong to $B(E, \mathfrak{C})$ (cf. lemma 3.26). We omit the proof of (*) if $n = -1$ and $n = 0$ since it is very similar to the proof of lemma 3.24.

Now let the relation (*) be true for any $n \geq 0$. Then

$$\begin{aligned}\psi^{n+1}(f')(x,d) &= \sum_{z \in E} (p(x,d,z) - \delta_{xz})\, v^{f'}_{n+1}(z) - v^{f'}_n(x)\\ &= \sum_{z \in E} (p(y,d,z) - \delta_{yz})\, v^{f'}_{n+1}(z) - v^{f'}_n(y)\\ &= \psi^{n+1}(f')(y,d)\end{aligned}$$

for $x,y \in E$ with $\varphi(x) = \varphi(y)$ and $d \in D(x)$.

Hence,

$$\psi^{n+1}(f')(y,d) = \psi^{n+1}(f')(x,d) \geq \psi^{n+1}(f')(x,d') = \psi^{n+1}(f')(y,d'),$$

for each $d \in D_{n+1}(x)$ and each $d' \in D_{n+1}(y)$, because we assumed that $D_{n+1}(y) \subset D_n(y) = D_n(x)$. Thus we conclude that $D_{n+1}(x) \subset D_{n+1}(y)$ and, consequently, $D_{n+1}(x) = D_{n+1}(y)$ by symmetry arguments. This identity implies the validity of (*) and show that

$$\bigcap_{x \in \varphi^{-1}(s)} D_n(x) \neq \emptyset$$

and $\mathbb{D}^{\mathfrak{C}}_n \neq \emptyset$. ∎

Theorem 3.30.

(i) Let $(V_{\mathfrak{C}}^*)$ be fulfilled. Then

$$\begin{array}{ccccc}\bar{\mathfrak{G}}_{N+1}(f) = \emptyset & \Longleftrightarrow & f^{\infty} & \text{is N-optimal on } \mathfrak{R}_S^{\prime} \\ \Updownarrow & & & \Updownarrow \\ \mathfrak{G}_{N+1}(\theta f) = \emptyset & \Longleftrightarrow & (\theta f)^{\infty} & \text{is N-optimal on } \mathfrak{R}_S^{\mathfrak{C}} .\end{array}$$

(ii) If (V^*) is fulfilled and if for some $f \in \mathfrak{D}'$ and $N \in \mathbb{N}_{-1}$

$$\mathfrak{G}_{N+1}(f) = \emptyset,$$

then $(\theta f)^{\infty}$ is an N-optimal policy on the class $\mathfrak{R}_S$.

Proof:

(i) If we mark the relations due to the diagram

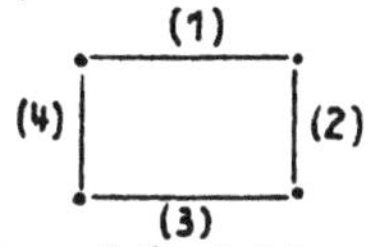

then (1) follows from theorem 2.40 (i), because $P(g)$ is quasi-compact for each $g \in \mathfrak{D}'$ (cf. lemma 1.28). The relation (3) follows from part (i) of the same lemma. The equivalence relation (2) results from theorem 3.11. (4) can be obtained by the following equivalence relations

$$\bar{\mathfrak{G}}_{N+1}(f) = \emptyset \Longleftrightarrow \max_{g \in \mathfrak{D}'} \bar{\Psi}^{N+1}(g,f) = 0 \Longleftrightarrow \max_{h \in \mathfrak{D}^{\mathfrak{C}}} \Psi^{N+1}(h,\theta f) = 0$$

$$\overset{(*)}{\Longleftrightarrow} \max_{h \in \mathfrak{D}} \Psi^{N+1}(h,\theta f) = 0 \Longleftrightarrow \mathfrak{G}_{N+1}(\theta f) = \emptyset,$$

where (*) follows from lemma 3.29.

(ii) $\mathfrak{G}_{N+1}(\theta f) = \emptyset$ due to (i). Since (V^*) is supposed to be fulfilled theorem 2.40 (ii) can be applied. ∎

Similarly to theorem 3.25 the following conclusion is obvious with respect to sensitive criteria. If the HOWARD algorithm can be applied within the class $\mathfrak{R}_S$ of the original model $\mathfrak{M}$ (i.e., if (V^*) is fulfilled), then it can applied within the class $\mathfrak{R}_S^{\prime}$ of the transformed model $\mathfrak{M}'$. If the algorithm breaks off with respect to $\mathfrak{M}'$, then we can find not only an N-optimal policy $\tau^* \in \mathfrak{R}_S$, but also an N-optimal policy on $\mathfrak{R}_S$ (instead of $\mathfrak{R}_S^{\mathfrak{C}}$) by $\hat{\theta}\tau^*$. (It is unknown whether this is true if only $(V_{\mathfrak{C}}^*)$ (instead of (V^*)) is fulfilled).

As shown above (section 2.4) the condition (V^*) is automatically fulfilled if E is finite. In this case the optimality assertions with respect to $\mathfrak{R}_S^{\prime}$ ($\mathfrak{R}_S$, respectively) can be immediately extended to the classes $\mathfrak{R}'$ ($\mathfrak{R}$, respectively).

If the state space is non-finite, then we "only" obtain

<u>Theorem 3.31.</u> Let (V**) (cf. section 2.6) be fulfilled. Then

$$\begin{array}{ccc} \mathcal{B}_{N+1}(f) = \emptyset & \Longleftrightarrow & f^{\infty} \text{ is N-optimal on } \mathcal{R}'_M \\ \Updownarrow & & \Uparrow \\ \mathcal{B}_{N+1}(\theta f) = \emptyset & \Longleftrightarrow & (\theta f)^{\infty} \text{ is N-optimal on } \mathcal{R}_M \end{array}$$

for $f \in \mathbb{D}'$ and $N \in \mathbb{N}_{-1}$.

<u>Proof:</u> If (V**) is fulfilled, then

$$\bar{P}(f)(s, \varphi(A_\varepsilon)) \geq 1 - \varepsilon$$

for a given $\varepsilon > 0$, the finite set $\varphi(A_\varepsilon)$, $s \in E'$, and $f \in \mathbb{D}'$. Thus theorem 2.49 can be applied to both $\mathfrak{M}$ and $\mathfrak{M}'$. (1) and (3) (with respect to the usual notation of such diagram) follow. Since (V**) implies (V*) (4) is a special case of theorem 3.30, and, finally, (2) is a consequence of theorem 3.11. ∎

3.5. DISCUSSION

Roughly speaking, the main result of the statements given so far consists in the fact that the solution of all Markovian decision problems considered with respect to a subclass of piecewise constant policies for the original model can be reduced to the solution of the related problems for the transformed model.

The optimal policies are determined by one-to-one relations. The different version of the HOWARD algorithms described here show a completely consistent mode of operation. Moreover, the exclusive consideration of piecewise constant policies with respect to the original model does not mean any restriction for a large class of Markovian decision problems, e.g., for finite stage problems,for discounted problems, and, under certain conditions, for problems concerning average reward and sensitive criteria.

Obviously, in many cases we can expect a substantial reduction of the effort if we use the transformed model. Since a solution of infinite problems is generally impossible in terms of figures the strong lumpation is a matter of principle for solving such problems.

The notion "effort" mentioned above is a very complex one. In order to analyse the effort we enter into the following aspects:
If we calculate within the transformed model, but not within the original model, then we can expect in view of the generally smaller "size" of the state space E':

(a) a smaller quantity of initial data,

(b) a reduction of the number of operations within the calculations, and

(c) a decrease of the required computer capacity.

Of course, we first have to check whether the original model allows a strong lumpation or not. This checking is also connected with an initial effort, which can hardly be investigated in a quantitative sense. However, the reductions mentioned in (a) and (b) are possible to be expressed by numerical calculations:

Let us assume that E and E' are finite sets with

$$e := \text{card } E' \text{ and } \alpha := e^{-1} \text{ card } E.$$

Furthermore, let us assume that

$$D(x) = D(y) =: D$$

for all $x,y \in E$ and $c :=$ card D.

Table 1 demonstrates the decrease of the number of initial data and of the numerical effort quantitatively.

Table 1

i	z_i	z_i'	$\lim_{e\to\infty} \frac{z_i'}{z_i}$
1	$c\alpha e(\alpha e+1)$	$ce(e+1)$	$\frac{1}{\alpha^2}$
2	$\alpha e(c(2\alpha e+1)-1)$	$e(c(2e+1)-1)$	$\frac{1}{\alpha^2}$
3	$\frac{\alpha^3 e^3}{3} + \alpha e(c(2\alpha e + 2)-1)$	$\frac{e^3}{3} + (c(2e+1)-1)$	$\frac{1}{\alpha^3}$

z_i and z_i', resp., are

for i = 1: numbers of initial data used within $\mathfrak{M}$ and $\mathfrak{M}'$, resp.

for i = 2: numbers of numerical operations for calculating V^*_{n+1} from V^*_n and $\overline{V}^*_{n+1}$ from $\overline{V}^*_n$, resp.

for i = 3: numbers of numerical operations within $\mathfrak{M}$ and $\mathfrak{M}'$, resp., for maximizing the discounted total reward.

There is to remark: We interpret the coefficients of c transition matrices of the size (e,e) and the c e-dimensional vectors of reward as initial data. The total size of all this data is ce(e+1).

Moreover, a policy improvement step in the case of a discounted total reward (with respect to the transformed model) consists in solving the two following problems:

- the calculation of $\overline{V}_{\beta}^{f}$ for a given $f \in \mathbb{D}'$ by solving the equation

$$(I - \beta\overline{P}(f))\overline{V}_{\beta}^{f} = \overline{r}(f),$$

- the following maximization of

$$\{\overline{r}(s,d) + \beta \sum_{s' \in E'} \overline{p}(s,d,s')\overline{V}_{\beta}(s')\}$$

  with respect to $d \in D$.

Here we assume that the solution of the linear equation system (by the Gaussian elimination procedure) requires about $\frac{1}{3}e^{3}$ numerical operations (basic arithmetic operations/maximization of two numbers). For "large" e it can be stated (cf. the last column of the table 1) that the quotient of the quantities of initial data and of the computational effort with respect to the transformed and the original model, respectively, are approximately equal to $1/\alpha^{2}$? This value approximately decreases to $1/\alpha^{3}$ in the case of discounted models. Without going into details we remark that one also finds the approximate value of $1/\alpha^{3}$ if the average reward maximization in the single case is considered. This quotient even becomes smaller if we maximize the average total reward in the multichain case or with respect to sensitive criteria, because the dimension of the linear equation systems to be solved will multiply.

The partially essential possibilities described above and bases on calculating within the transformed model can be used if the original model allows a strong lumpation. This condition is a very restrictive one (cf. SMITH [107]). Here is the critical disadvantage of this method. Moreover, in some cases it seems to be difficult to decide whether a strong lumpation is possible and which decomposition can be used if such a strong lumpation is possible.

Nevertheless, it seems to be justified to expect that there is a lot of practical problems where the decomposition into cluster sets is given in a natural way. For instance, we focus our attention on systems whose behaviour does not essentially depend on the behaviour of the single elements but more strongly on the functions and connections of certain sub-systems. An example for such a system was given by DIETZ [24] who considered a maintaining problem of traffic lights.

Furthermore, a very interesting question is whether a (modified) strong lumpation can be advantageous for solving MDP's if the original model does not fulfil the strong lumpability condition. The answer is in the affirmative if the initial model can be transformed into a strongly lumpable one by "small" disturbances as was shown in [24]. If, for instance, the underlying Markov chains are strongly lumpable but the pay-off function not being piecewise constant, then bounds for the reward functions of the initial model can be obtained (Remark that the substitution of the initial pay-off function by the arithmetic means with respect to the cluster sets is very unfavourable in some cases, cf. the example in [24]).

On the other hand, if the strong lumpability conditions **with respect to the initial model are fulfilled except** for the lumpability condition concerning the underlying Markov chain, then a strongly lumpable MDM can be obtained by "small" disturbances of the transition law in some cases. (Related disturbance problems were considered by KOROLJUK and TURBIN [73], SCHWEITZER [103], and TAKAHASHI [109].) Here we only cite the following

Lemma 3.32. Let $e := \text{card } E < \infty$ and let $\varepsilon > 0$ be a sufficient small number with

$$(1+\varepsilon)^{-1}P(h) \leqq \tilde{P}(h) \leqq (1+\varepsilon)P(h)$$

for all $d \in D$. Then

$$|\tilde{g}^* - g^*| \leqq 2\varepsilon(e-1)g^* + O(\varepsilon^2)\mathbf{1} .$$

The proof of this lemma bases on a lemma given by TAKAHASHI [109].

Finally, we remark that there are several publications which refer to different lumpation procedures in order to solve MDP's approximately (see for example BERTSEKAS [7], FOX [35], [36], HAHNEWALD-BUSCH and NOLLAU [40], [41], [86], PORTEUS [90] and SMITH [107]). The method developed here differs from the others with respect to two facts: First, only one "new" model is considered and second, the solution of problems within the "new" model provides an exact solution of certain (sub-)problems within the original model.

APPENDIX: MATRICES AND LINEAR OPERATORS

A 1. MATRICES

Definition A 1. Let E and E' be non-empty countable sets. A function $M: E \times E' \to \mathbb{C}$ is called a (complex) $E \times E'$-matrix (respectively: matrix).
If $M(E \times E') \subset \mathbb{R}$, then M is called a real matrix. If E and E' are finite sets, then M is called a finite matrix. For $E = E'$ the Matrix M is called a quadratic matrix. If card $E = 1$, then M denotes an E'-row vector and if card $E' = 1$, M denotes an E-column vector.

The values $M(x,y)$, $x \in E$, $y \in E'$, are said to be the elements of M. For non-empty sets $A \subset E$ and $A' \subset E'$ let $M_{AA'} := M|_{A \times A'}$.
$M_{AA'}$ is called a submatrix of M. $M_{\{x\}E'}$ and $M_{E\{y\}}$ $(x \in E,\ y \in E')$, respectively, are special submatrices of M, the so-called x-row and y-column, respectively.

The introduced notations represent generalizations of the corresponding notations as used in the well-known theory of finite matrices. (Let us remark that in the finite case a matrix is a rectangle scheme of numbers, i.e., the table of values of the matrix in the sense of the definition A1. In the countable case this approach is not suitable. As far as possible we try to identify a matrix with its table of values.) Analogously to the finite case the set $\mathbb{C}^{E \times E'}$ is supplied with an addition and a scalar multiplication:
The matrix $M_1 + M_2$ defined by

$$(M_1 + M_2)(x,y) := M_1(x,y) + M_2(x,y),$$

$(x,y) \in E \times E'$, is the so-called sum of M_1 and M_2. Further let

$$(cM)(x,y) := cM(x,y)$$

for $c \in \mathbb{C}$ and $(x,y) \in E \times E'$. With respect to these operations $\mathbb{C}^{E \times E'}$ represents a complex vector space, whose zero is the zero matrix

$$O(x,y) := 0 \qquad ((x,y) \in E \times E').$$

The definition of matrix multiplication is based on the following notion of convergence.

Definition A 2. Let E be a countable set and $f: E \to \mathbb{C}$. If for every bijection $\lambda: \mathbb{N} \to E$ the limit

$$c := \lim_{n \to \infty} \sum_{k=1}^{n} f(\lambda(k))$$

exists and if this limit is independent of the choice of λ, we write

$$\sum_{x \in E} f(x) := c.$$

Then the series $\sum_{x\in E} f(x)$ is said to be convergent.

Let us remark that this concept of convergence does not correspond to the usual concept of convergence for $E = \mathbb{N}$. The series $\sum_{x\in E} f(x)$ is convergent, if and only if $\sum_{n=1}^{\infty} f(n)$ absolutely converges in the usual sense.

<u>Definition A 3.</u> Let M and N be an $E \times E'$-matrix and an $E' \times E''$-matrix, respectively. If the series

$$\sum_{z\in E'} M(x,z)N(z,y)$$

converges for all $x \in E$ and all $y \in E'$, then the $E \times E''$-matrix MN defined by

$$(MN)(x,y) := \sum_{z\in E'} M(x,z)N(z,y), \qquad (x,y) \in E \times E'',$$

is called the product of M and N.

For a matrix $M \in \mathbb{C}^{E\times E'}$ let $|M|$ be the matrix with the elements

$$|M|(x,y) := |M(x,y)|$$

($(x,y) \in E \times E'$). One immediately obtains the

<u>Lemma A 1.</u> The product MN of an $E \times E'$-matrix M and an $E \times E''$-matrix N exists, if and only if there exists $|M||N|$.

<u>Lemma A 2.</u> Let E, E' and E'' be non-empty countable sets, let K and L be $E \times E'$-matrices, and let M and N be $E'\times E''$-matrices.

(i) If LM and LN exist, then the product L(M + N) exists, too, and $L(M + N) = LM + LN$.

(ii) If KM and LM exist, then so does the product (K + L)M, and $(K + L)M = KM + LM$.

<u>Proof:</u> (i) Due to lemma A1 one obtains for any $x\in E$, $y\in E''$ and any bijection $\lambda: N \to E'$:

$$\begin{aligned}
&\sum_{n=1}^{\infty} |L(x,\lambda(n))|\ |(M+N)(\lambda(n),y)| \\
&\le \sum_{n=1}^{\infty} (|L(x,\lambda(n))|\ |M(\lambda(n),y)| + |L(x,\lambda(n))||N(\lambda(n),y)|) \\
&= |L||M|(x,y) + |L||N|(x,y) < \infty .
\end{aligned}$$

Thus the existence of L(M + N) is proved. We then obtain for all $x \in E$, $y \in E''$:

$$\begin{aligned}
(L(M + N))(x,y) &= \sum_{z\in E'} L(x,z)((M + N)(z,y)) \\
&= \sum_{z\in E'} L(x,z)M(z,y) + \sum_{z\in E'} L(x,z)N(z,y)
\end{aligned}$$

$= (LM)(x,y) + (LN)(x,y).$

(ii) This proposition can be proved in a similar way. ∎

The associative law does not hold in general. I.e., the equation $L(MN) = (LM)N$ need not be true even if the products on both sides do exist. However, we have the proposition

Lemma A 3. Let E_i $(i=1,\dots,4)$ be non-empty countable sets, and let L, M, and N be $E_1 \times E_2$-, $E_2 \times E_3$-, and $E_3 \times E_4$-matrices, respectively.

(i) If L, M, and N are non-negative matrices and if $L(MN)$ and $(LM)N$ exist, then $L(MN) = (LM)N$.

(ii) If there exist $|L|(|M||N|)$ and $(|L||M|)|N|$, then $L(MN) = (LM)N$.

Proof:

(i) One obtains

$$(L(MN))(x,y) = \sum_{n=1}^{\infty} L(x,\lambda(n))(MN)(\lambda(n),y) = \sum_{n=1}^{\infty} \sum_{m=1}^{\infty} L(x,\lambda(n))M(\lambda(n),\mu(m))N(\mu(m),y)$$

for any $x \in E_1$, $y \in E_4$ and any bijections $\lambda: \mathbb{N} \to E_2$ and $\mu: \mathbb{N} \to E_3$. Since this double series absolutely converges we obtain the equations

$$(L(MN))(x,y) = \sum_{m=1}^{\infty} \sum_{n=1}^{\infty} L(x,\lambda(n))M(\lambda(n), \mu(m))N(\mu(m),y) = \sum_{m=1}^{\infty} (LM)(x,\mu(M))N(\mu(m),y) = ((LM)N)(x,y).$$

(ii) Using part (i) and lemma A 2 this proposition can be proved by decomposing the matrices L, M, and N into the real and imaginary parts, which are then decomposed into the positive and negative parts. ∎

Now let us consider the special case of quadratic matrices and single out a subclass, within which we can calculate as in the algebra of complex numbers. The set of all $E \times E$-matrices M which fulfil the condition

$$\|M\| := \sup_{x \in E} \sum_{y \in E} |M(x,y)| < \infty$$

will be denoted by $\mathcal{R}(E)$.

Lemma A 4. $\mathcal{R}(E)$ forms an algebra over $\mathbb{C}$ with respect to addition, matrix multiplication, and multiplication with a complex number with the 1-element I:

$$I(x,y) := \delta_{xy} \quad (x,y \in E)$$

($\delta_{..}$-Kronecker-symbol), which is called the unit matrix. The function $M \to \|M\|$ is a norm on $\mathfrak{R}(E)$. $\mathfrak{R}(E)$ forms a Banach algebra with regard to this norm.

The <u>proof</u> is obvious.

<u>Note:</u> Let $M \in \mathbb{C}^{E \times E}$. A matrix M' is called an inverse of M if $M'M$ and MM' exist and if $MM' = M'M = I$.
If $M' \in \mathfrak{R}(E)$ is an inverse of $M \in \mathfrak{R}(E)$, then this inverse is unique and it will be denoted by M^{-1}.

Furthermore, we can introduce the so-called lexicographical order with respect to real $E \times E'$-matrices, whose second index set E' has the form $E' = \{k \in Z: k \geq k_0\}$ with a suitable $k_0 \in Z$ (Z - set of all integers). We say that such a matrix is lexicographically non-negative ($M \succcurlyeq 0$) if the first non-zero element of each row of M is non-negative.
We define

$$M \succ 0 :\iff M \succcurlyeq 0 \wedge M \neq 0$$

and

$$M \succcurlyeq M' :\iff M - M' \succcurlyeq 0.$$

A 2. <u>LINEAR OPERATORS</u>

Let B and B' be complex Banach spaces. A mapping $T: B \to B'$ is linear if

$$T(a_1x_1 + a_2x_2) = a_1T(x_1) + a_2T(x_2)$$

for all $a_1, a_2 \in \mathbb{C}$ and $x_1, x_2 \in B$. Such a mapping T is called bounded if

$$\|T\| := \sup_{x:\|x\|_B \leq 1} \|Tx\|_{B'} < \infty,$$

where $\|\cdot\|_B$ and $\|\cdot\|_{B'}$ are the norms on B and B', respectively.
The set of all bounded linear operators $T: B \to B'$ is denoted by $\mathfrak{L}(B,B')$.
If $B = B'$, then $\mathfrak{L}(B)$ means $\mathfrak{L}(B,B')$.

We define an addition and a scalar multiplication on $\mathfrak{L}(B,B')$ in the following way:

$$(T_1 + T_2)(x) := T_1(x) + T_2(x) \qquad (x \in B),$$
$$(aT_1)(x) := aT_1(x) \qquad (a \in \mathbb{C},\ x \in B)$$

for $T_1, T_2 \in \mathfrak{L}(B,B')$.
If $T: B \to B'$ and $S: B' \to B$ are linear mappings, then the product ST is defined by

$$(ST)(x) := (S \circ T)(x) = S(T(x)) \qquad (x \in B).$$

Lemma A 5. ([70]) Let B and B' be complex Banach spaces. If $\mathfrak{L}(B,B')$ is supplied with the addition and scalar multiplication defined above, it is a complex Banach space with the norm $\|\cdot\|$. In particular, $\mathfrak{L}(B)$ is a complex Banach algebra supplied with the product introduced above and the unit element I given by $Ix := x$ – the identity operator.

If a linear operator T represents a one-to-one mapping of B onto B, then there exists a bounded linear operator (by a well-known BANACH theorem), which satisfies $TT^{-1} = T^{-1}T = I$. This inverse T^{-1} is uniquely determined.

The set $\ker T := \{x \in B: Tx = 0\}$ is called the kernel or the null space of T. Obviously, this set is a closed subspace of B. One easily obtains the relations

$$\ker T \subset \ker T^2 \subset \dots \subset \ker T^n \subset \dots$$

If there is a positive integer n_0 with $\ker T^{n_0} = \ker T^{n_0+1}$, then n_0 is called the index of T. Otherwise, T has an infinite index.

Let B be a Banach space and $T \in \mathfrak{L}(B)$. The set $\rho(T)$ of all complex numbers λ, for which $T - \lambda I$ is non-singular, is called the resolvent set of T. All complex numbers λ with $\lambda \notin \rho(T)$ form a non-empty closed set $\sigma(T)$ called the spectrum of T. The mapping

$$\lambda \to R(\lambda,T) := (T - \lambda I)^{-1} \qquad (\lambda \in \rho(T))$$

is called the resolvent of T [1]).

If $\lambda \in \sigma(T)$ has the property

$$\ker(T - \lambda I) \neq \{0\},$$

then λ is called an eigenvalue of T, and the dimension of $\ker(T - \lambda I)$ is called the (geometric) multiplicity of the eigenvalue λ. The index $T - \lambda I$ is also called the index of λ. $\operatorname{spr} T := \lim\limits_{n\to\infty} \sqrt[n]{\|T\|^n}$ is said to be the spectral radius of T.

Lemma A 6. If T is a bounded linear operator on a complex Banach space B, then

(i) $\operatorname{spr} T = \sup\{|\lambda| : \lambda \in \sigma(T)\}$,

(ii) $\sigma(T)$ is bounded and non-empty,

(iii) $\operatorname{spr} T \leqq \|T\|$,

(iv) the resolvent is given by the series

$$R(\lambda;T) = -\lambda^{-1} \sum_{n=0}^{\infty} \lambda^{-n} T^n$$

[1]) This definition was given by KATO [70]. It differs from the definition due to DUNFORD and SCHWARTZ [28] by (-1).

(which converges with respect to the norm topology) for all $\lambda \in \mathbb{C}$ with $|\lambda| > \operatorname{spr} T$, and

(v) each eigenvalue λ of T with $|\lambda| = \|T\|$ has the index 1.

<u>Proof:</u> (i) - (iv): see [70], (v): see [123]. ∎

$R(\cdot,T)$ is a holomorphic function on $\rho(T)$ due the proposition (iv) of the preceding lemma. An isolated point $\lambda_0 \in \sigma(T)$ is called a pole of the order n if $R(\lambda,T)$ has a pole of the order n for $\lambda = \lambda_0$.

<u>Lemma A 7.</u> A pole of the order n is an eigenvalue of T with the index n.

<u>Proof:</u> [28], theorem VII. 3.18. ∎

Let σ be a subset of $\sigma(T)$. This subset σ is called a spectral set if σ is open as well as closed in $\sigma(T)$. In this case a rectiable Jordan curve Γ containing the spectrum of T can be given. Furthermore, there is a complex-valued function which is analytic on a neighbourhood of the set of inner points of Γ with

$$f(x) = \begin{cases} 1 & x \in \sigma \\ 0 & x \in \sigma(T)\setminus\sigma \end{cases} .$$

Then there exists the integral

$$e_\sigma(T) := -\frac{1}{2\pi i}\int_\Gamma f(\lambda)R(\lambda;T)\,d\lambda .$$

This integral defines a bounded linear operator $e_\sigma(T)$ with $(e_\sigma(T))^2 = e_\sigma(T)$. (I.e., $e_\sigma(T)$ is a projection).

<u>Theorem A 8.</u> (Spectral decomposition theorem)

(i) If $B_\sigma := e_\sigma(T)B$ and $B'_\sigma := (I - e_\sigma(T))B$, then B is the direct sum of B_σ and B'_σ:

$$B = B_\sigma \oplus B'_\sigma .$$

(ii) $TB_\sigma \subset B_\sigma$ and $TB'_\sigma \subset B'_\sigma$.

(iii) Let T_σ be the restriction of T on B_σ and T'_σ the restriction of T on B'_σ, then

$$\sigma(T_\sigma) = \sigma \text{ and } \sigma(T'_\sigma) = \sigma(T)\setminus\sigma.$$

<u>Proof:</u> [70], theorem III. 6.17. ∎

If λ is an eigenvalue as well as an isolated point of the spectrum of T, then we write $e_\lambda(T)$ and T_λ instead of $e_{\{\lambda\}}(T)$ and $T_{\{\lambda\}}$, respectively.

Theorem A 9. If λ is a pole (of the order n) of the operator T, then

$$B_\lambda = \ker(T - \lambda I)^n \text{ and } B_\lambda' = (T - \lambda I)^n B.$$

Proof: [28], theorem VII. 3.24. ∎

Finally, we state the following

Lemma A 10. Let σ be a spectral set of the operator $T \in \mathfrak{L}(B)$. Then

$$\sigma(e_\sigma(T)) = \{0, 1\}.$$

Proof: Using the function f introduced above (lemma A 7) we obtain

$$\sigma(e_\sigma(T)) = f(\sigma(T)) = f(\sigma) \cup f(\sigma(T) \setminus \sigma) = \{0, 1\}$$

in view of the spectral mapping theorem (cf. [28], theorem VII. 3.11). ∎

A 3. REPRESENTATION OF LINEAR OPERATORS BY MATRICES

Let E be a countable set and B(E) the set of all bounded complex-valued functions defined on E. Then any matrix $M \in \mathfrak{R}(E)$ generates a linear operator τ_M: B(E) → B(E) by

$$\tau_M f(x) := \sum_{y \in E} M(x,y) f(y) \qquad (x \in E,\ f \in B(E))$$

with the property that

$$|\tau_M f(x)| \leq \sum_{y \in E} |M(x,y)|\ |f(y)| \leq \|f\| \sum_{y \in E} |M(x,y)|$$

$$\leq \|f\|\ \|M\| < \infty,$$

i.e., $\|\tau_M\| \leq \|M\|$.

Let us remark that the one-to-one mapping $\tau: M \to \tau_M$ is compatible with linear operations and the multiplication as can easily be seen. Thus we obtain

Lemma A 11. The mapping T is a continuous isomorphism of the Banach algebra $\mathfrak{R}(E)$ onto a closed subalgebra of $\mathfrak{L}(B(E))$ transforming the unit matrix into the identity operator.

REFERENCES

[1] ASO, H.; KIMURA, M.: An application of Markov potential theory to Markovian decision processes.
Internat. J. Systems Sci. 4 (1973), 907 - 932.

[2] BATHER, J.: Optimal decision procedures for finite Markov chains. Parts I - III.
Adv. Appl. Prob. 5 (1973), 328 - 339, 521 - 540, 541 - 553.

[3] BATHER, J.: Optimal stationary policies for denumerable Markov chains in continuous time.
Adv. Appl. Prob. 8 (1976), 144-158.

[4] BAUER, H.: Wahrscheinlichkeitstheorie und Grundzüge der Maßtheorie.
Walter de Gruyter & Co., Berlin 1968.

[5] BELLMAN, R.: A Markovian decision process.
J. Math. Mech. 6 (1957), 679 - 684.

[6] BELLMAN, R.: Dynamic Programming.
Princeton Univ. Press, Princeton 1957.

[7] BERTSEKAS, D.P.: Convergence of discretization procedures in dynamic programming.
IEEE Trans. Automat. Control, June 1975, 415 - 419.

[8] BLACKWELL, D.: Discounted dynamic programming.
Ann. Math. Statist. 36 (1965), 226 - 235.

[9] BLACKWELL, D.: Discrete dynamic programming.
Ann. Math. Statist. 33 (1962), 719 - 726.

[10] BROWN, B.W.: On the iterative method of dynamic programming on a finite space discrete time Markov process.
Ann. Math. Statist. 36 (1965), 1279 - 1285.

[11] CHUNG, K.L.: Markov Chains with Stationary Transition Probabilities.
Springer-Verlag New York, Heidelberg, Berlin 1967 (2. ed.).

[12] COOKE, R.G.: Infinite Matrices and Sequence Spaces.
MacMillan & Co., London 1950.

[13] DE GHELLINCK, G.T.; EPPEN, G.D.: Linear programming solutions for separable Markovian decision problems.
Management Sci. 13 (1967), 371 - 394.

[14] DENARDO, E.V.: A Markov decision problem.
In: Mathematical Programming, 33 - 68. Math. Res. Center Publ., No. 30, Academic Press, New York 1973.

[15] DENARDO, E.V.: Computing a bias-optimal policy in a discrete-time Markov decision problem.
Operations Res. 18 (1970), 279 - 289.

[16] DENARDO, E.V.: Separable Markovian decision problems.
Management Sci. 14 (1968), 451 - 462.

[17] DENARDO, E.V.; MILLER, B.L.: An optimality condition for discrete dynamic programming with no discounting.
Ann. Math. Statist. 39 (1968), 1220 - 1227.

[18] DERMAN, C.: Denumerable state Markovian decision processes - average cost criterion.
Ann. Math. Statist. 37 (1966), 1545 - 1554.

[19] DERMAN, C.: Finite State Markovian Decision Processes. Academic Press, New York 1970.

[20] DERMAN, C.: Markovian sequential control processes - denumerable state space. J. Math. Anal. Appl. 10 (1965), 295 - 302.

[21] DERMAN, C.; LIEBERMAN, G.J.: A Markovian decision model for a joint replacement and stocking problem. Management Sci. 13 (1966), 609 - 617.

[22] DERMAN, C.; STRAUCH, R.E.: A note on memoryless rules for controlling sequential control processes. Ann. Math. Statist. 37 (1966), 276 - 278.

[23] DERMAN, C.; VEINOTT, A.F., Jr.: A solution to a countable system of equations arising in Markovian decision processes. Ann. Math. Statist. 38 (1967), 582 - 584.

[24] DIETZ, H.-M.: Untersuchungen zu Markovschen Entscheidungsprozessen mit abzählbarem Zustandsraum, Dissertation, Technische Universität Dresden, 1979.

[25] DIETZ, H.-M.; NOLLAU, V.: Schwache Vergröberungen für endliche Markovsche Ketten. Wiss. Zeitschr. Techn. Univ. Dresden 27 (1978), 349 - 353.

[26] DIEUDONNE, J.: Grundzüge der modernen Analysis. Bd. 2. Deutscher Verlag der Wissenschaften, Berlin 1975.

[27] DOSHI, B.T.: Continuous time control of a Markov process on an arbitrary state space: average return criterion. Stochastic Processes Appl. 4 (1976), 55 - 77.

[28] DUNFORD, N.; SCHWARTZ, J.T.: Linear Operators. Part I: General Theory. Interscience, New York 1958, Vol. I.

[29] DYNKIN, E.B.: Markov Processes. Springer Verlag, Berlin 1965.

[30] FEINBERG, E.A.: Controlled Markov Processes with a finite number of states and with compact sets of policies (Russ.). Teor. Verojatnost. i Prim. XX (1975), 873 - 880.

[31] FEDERGRUEN, A.; SCHWEITZER, P.J.; TIJMS, H.C.: Valueiteration in undiscounted Markov decision problems. Parts I - III. In: Markov Decision Theory. Eds. H.C. Tijms & J. Wessels. Math. Centre Tracts No. 93, pp. 119 - 140, 141 - 151, 153 - 160. Amsterdam 1977.

[32] FISHER, L.; ROSS, S.M.: An example in denumerable decision processes. Ann. Math. Statist. 39 (1968), 674 - 675.

[33] FLYNN, J.: Averaging versus discounting in dynamic programming: a counterexample. Ann. Statist. 2 (1974), 411 - 413.

[34] FLYNN, J.: Conditions for the equivalence of optimality criteria in dynamic programming. Ann. Statist. 4 (1976), 936 - 953.

[35] FOX, B.L.: Discretizing dynamic programs. J. Optimization Theory and Appl. 11 (1973), 228 - 234.

[36] FOX, B.L.: Finite-state approximation to denumerable state dynamic programming.
J. Math. Anal. Appl. 34 (1971), 665 - 670.

[37] FURUKAWA, N.: Markovian decision processes with compact action spaces.
Ann. Math. Statist. 43 (1972), 1612 - 1622.

[38] GRINOLD, R.C.: Elimination of suboptimal actions in Markov decision problems.
Operations Res. 21 (1973), 848 - 851.

[39] GUBENKO, L.G.; STATLAND, E.S.: Controlled Markov Processes with discrete time parameter (Russ.).
Teor. Verojatnost. i Mat. Statist. 7 (1972), 51 - 64.

[40] HAHNEWALD-BUSCH, A.: Approximative Methoden zur Lösung von Steueraufgaben bei Markov'schen Ketten.
Dissertation, Technische Universität Dresden, 1976.

[41] HAHNEWALD-BUSCH, A.; NOLLAU, V.: An approximation procedure for stochastic dynamic programming in countable state space.
MOS, Ser. Opt. 9 (1978), 100 - 117.

[42] HAHNEWALD-BUSCH, A.; NOLLAU, V.: A Method of Clustering for discounted Markovian Decision Problems.
MOS, Ser. Opt. 12 (1981), 137 - 147.

[43] HARRISON, J.M.: Discrete dynamic programming with unbounded rewards.
Ann. Math. Statist. 43 (1972), 636 - 644.

[44] HASTINGS, N.A.J.: A test for nonoptimal actions in undiscounted finite Markov decision chains.
Management Sci. 23 (1976), 87 - 92.

[45] HASTINGS, N.A.J.: Bounds on the gain of a Markov decision process.
Operations Res. 19 (1971), 240 - 243.

[46] HASTINGS, N.A.J.; MELLO, J.M.C.: Tests for suboptimal actions in discounted Markov programming.
Management Sci. 19 (1972/73), 1019 - 1022.

[47] HASTINGS, N.A.J.; VAN NUNEN, J.A.E.E.: The action eliminating algorithm for Markov decision processes.
In: Markov Decision Theory. Eds. H.C. Tijms & J. Wessels.
Math. Centre Tracts No. 93, pp. 161 - 170. Amsterdam 1977.

[48] HINDERER, K.: Estimates for finite-stage dynamic programs.
J. Math. Anal. Appl. 55 (1976), 207 - 238.

[49] HINDERER, K.: Foundations of Non-stationary Dynamic Programming with Discrete Time Parameter.
Lecture Notes in Operations Research and Mathematical Systems No. 33. Springer-Verlag, Berlin - Heidelberg - New York 1970.

[50] HINDERER, K.: Neuere Resultate in der Stochastischen Dynamischen Optimierung.
ZAMM 55 (1975), T16 - T26.

[51] HINDERER, K.: On approximate solutions of finite-stage dynamic programs.
Paper presented at the International Conference on Dynamic Programming, University of British Columbia, Vancouver 1977.

[52] HINDERER, K.; HÜBNER, G.: An improvement of J.F. SHAPIRO's turnpike-theorem for the horizon of finite stage, discrete dynamic programs.
Trans. 7th Prague Conference on Information Theory, Statistical Decision Functions, and Random Processes. Vol. A, pp. 245 - 255. Prague 1974.

[53] HINDERER, K.; HÜBNER, G.: On approximate and exact solutions for finite stage dynamic programs.
In: Markov Decision Theory. Eds. H.C. Tijms & J. Wessels. Math. Centre Tracts No. 93, pp. 57 - 76. Amsterdam 1977.

[54] HORDIJK, A.: A sufficient condition for the existence of an optimal policy with respect to the average cost criterion in Markovian decision processes.
Trans. 6th Prague Conference on Information Theory, Statistical Decision Functions, and Random Processes, pp. 263 - 274, Prague 1973.

[55] HORDIJK, A.: Dynamic Programming and Markov Potential Theory. Math. Centre Tracts No. 51, Mathematisch Centrum, Amsterdam 1974.

[56] HORDIJK, A.: On the convergence of average expected return in dynamic programming.
J. Math. Anal. Appl. 46 (1974), 542 - 544.

[57] HORDIJK, A.; SCHWEITZER, P.J.; TIJMS, H.C.: The asymptotic behaviour of the minimal total expected cost for the denumerable state Markov decision model.
J. Appl. Probability 12 (1975), 298 - 305.

[58] HORDIJK, A.; SLADKY, K.: Sensitive optimality criteria in countable state dynamic programming.
Mathematics of Operations Research 2 (1977), 1 - 14.

[59] HORDIJK, A.; TIJMS, H.C.: A counterexample in discounted dynamic programming.
J. Math. Anal. Appl. 39 (1972), 455 - 457.

[60] HOWARD, R.A.: Dynamic Programming and Markov Processes. MIT Press, Cambridge 1960.

[61] HÜBNER, G.: Improved procedures for eliminating suboptimal actions in Markov programming by the use of contraction properties.
Trans. 7th Prague Conference on Information Theory, Statistical Decision Functions, and Random Processes, Vol. A, pp. 257 - 263, Prague 1974.

[62] HÜBNER, G.: On the fixed points of the optimal reward operator in stochastic dynamic programming with discount factor greater than one.
ZAMM 57 (1977), 477 - 480.

[63] IWAMOTO, S.: Average reward Markovian decision processes in the completely ergodic case.
Bull. Math. Statist. 15 (1973), 55 - 68.

[64] JAQUETTE, S.C.: Markov decision processes with a new optimality criterion: discrete time.
Ann. Statist. 1 (1973), 496 - 505.

[65] JAQUETTE, S.C.: Markov decision processes with a new optimality criterion: small interest rates.
Ann. Math. Statist. 43 (1972), 1894 - 1901.

[66] KADOTA, Y.: Countable State Markovian Decision Processes under the DOEBLIN-Condition.
Res. Assoc. Statist. Sci. 19, 85 - 94.

[67] KAKUMANU, P.: Continuous time Markovian decision processes - average return criterion.
J. Math. Anal. Appl. 52 (1975), 173 - 188.

[68] KAKUMANU, P.: Nondiscounted continuous time Markovian decision processes with countable state space.
SIAM J. Control 10 (1972), 210 - 220.

[69] KALYMON, B.A.: Structured Markovian decision problems.
Naval Res. Logist. Quart. 20 (1973), 1 - 11.

[70] KATO, T.: Perturbation Theory for Linear Operators.
Springer Berlin 1966.

[71] KEMENY, J.G.; SNELL, J.L.: Finite Markov Chains.
(2.ed.) Van Nostrand, Princeton, 1976.

[72] KEMENY, J.G.; SNELL, J.L.; KNAPP, A.W.: Denumerable Markov Chains.
(2. ed.) Springer-Verlag New York - Heidelberg - Berlin 1976.

[73] KOROLJUK, V.S.; TURBIN, A.F.: Perturbation of operators on the spectrum and asymptotic expansion of Markov chains, (Ukrain.).
Dopovidi Akad. Nauk Ukrain. RSR Ser. A 1975, no. 5, 402 - 405.

[74] KÜENLE, H.-U.: Markovsche Entscheidungsmodelle mit allgemeinem Zustands- und Aktionsraum und ihre Anwendung in der Lagerhaltungstheorie.
Dissertation, Karl-Marx-Universität Leipzig, 1975.

[75] KUSHNER, H.J.: Introduction to Stochastic Control.
Holt, Rinehart and Winston, New York 1971.

[76] LIPPMAN, S.A.: Criterion equivalence in discrete dynamic programming.
Operations Res. 17 (1969), 920 - 923.

[77] LIPPMAN, S.A.: Maximal average-reward policies for semi-Markov decision processes with arbitrary state and action space.
Ann. Math. Statist. 42 (1971), 1717 - 1726.

[78] MAC QUEEN, J.: A modified dynamic programming method for Markovian decision problems.
J. Math. Anal. Appl. 14 (1966), 38 - 43.

[79] MAC QUEEN, J.: A test for suboptimal actions in Markovian decision problems.
Operations Res. 15 (1967), 559 - 561.

[80] MAITRA, A.: A note on undiscounted dynamic programming.
Ann. Math. Statist. 37 (1966), 1042 - 1044.

[81] MAITRA, A.: Discounted dynamic programming on compact metric spaces.
Sankhya, Ser. A, 30 (1968), 211 - 216.

[82] MAITRA, A.: Dynamic programming for countable state systems.
Sankhya, Ser. A, 27 (1965), 259 - 266.

[83] MANDL, P.: On the variance in controlled Markov chains.
Kybernetika (Praha) 7 (1971), 1 - 12.

[84] MILLER, B.L.; VEINOTT, A.F., Jr.: Discrete dynamic programming with a small interest rate.
Ann. Math. Statist. 40 (1969), 366 - 370.

[85] NEVEU, J.: Mathematische Grundlagen der Wahrscheinlichkeitstheorie. Springer-Verlag, München, Wien 1969

[86] NOLLAU, V.; HAHNEWALD-BUSCH, A.: An approximation procedure for stochastic dynamic programming based on clustering of state and action spaces.
MOS, Ser. Opt. 10 (1979), 121-130.

[87] ODONI, A.R.: On finding the maximal gain for Markov decision processes.
Operations Res. 17 (1969), 857 - 860.

[88] PITTEL, B.G.: Asymptotic estimations in dynamic problems (Russ.). Teor. Verojatnost. i Prim. XIV (1969), 250 - 268.

[89] PORTEUS, E.L.: Bounds and transformations for discounted finite Markov chains.
Operations Res. 23 (1975), 761 - 784.

[90] PORTEUS, E.L.: Some bounds for discounted sequential decision processes.
Management Sci. 18 (1971), 7 - 11.

[91] QUELLE, G.: Dynamic programming of expectation and variance. J. Math. Anal. Appl. 55 (1976), 239 - 252.

[92] RIEDER, U.: On HOWARD's policy improvement method.
Math. Operationsforsch. Statist. Ser. Optimization 8 (1977), 227 - 236.

[93] ROSENBLATT, M.: Functions of Markov processes that are Markovian. J. Math. Mech. 8 (1959), 585 - 596.

[94] ROSENBLATT, M.: Note correcting a remark in a paper of Karl Bosch. Z. Wahrscheinlichkeitstheorie und Verw. Gebiete 33 (1975), 219.

[95] ROSS, S.M.: Applied Probability Models with Optimization Applications.
Holden-Day, San Francisco 1970.

[96] ROSS, S.M.: Arbitrary state Markov decision processes.
Ann. Math. Statist. 39 (1968), 2118 - 2122.

[97] ROSS, S.M.: Non-discounted denumerable Markovian decision models. Ann. Math. Statist. 39 (1968), 412 - 423.

[98] ROSS, S.M.: On the nonexistence of ε-optimal randomized stationary policies in average cost Markov decision models.
Ann. Math. Statist. 42 (1971), 1767 - 1768.

[99] ROTHBLUM, U.G. Normalized Markov decision chains.
Operations Res. 23 (1975), 785 - 795.

[100] SCHÄL, M.: Conditions for optimality in dynamic programming and for the limit of n-stage optimal policies to be optimal.
Z. Wahrscheinlichkeitstheorie und Verw. Gebiete 32 (1975), 179 - 196.

[101] SCHÄL, M.: Ein verallgemeinertes stationäres Entscheidungsmodell der dynamischen Optimierung.
Operations Research-Verfahren X (1971), 145 - 171.

[102] SCHÄL, M.: On negative dynamic programming with irreducible Markov chains and the average cost criterion.
In: Dynamische Optimierung. Bonner Math. Schriften Nr. 98, pp. 93 - 97, Bonn 1977.

[103] SCHWEITZER, P.J.: Perturbation theory and finite Markov chains. J. Appl. Prob. 5 (1968), 403 - 413.

[104] SHAPIRO, J.F.: Turnpike planning horizons for a Markovian decision model.
Management Sci. 14 (1968), 292 - 300.

[105] SLADKY, K.: Necessary and sufficient optimality conditions for average reward of controlled Markov chains.
Kybernetika (Praha) 9 (1973), 124 - 137.

[106] SLADKY, K.: On the set of optimal controls for Markov chains with rewards.
Kybernetika (Praha) 10 (1974), 350 - 367.

[107] SMITH, J.L.: Markov decisions on a partitioned state space. IEEE Trans. Systems, Man, and Cybernet. SMC-1 (1971), 55 - 60.

[108] STRAUCH, R.E.: Negative dynamic programming. Ann. Math. Statist. 37 (1966), 871 - 890.

[109] TAKAHASHI, Y.: On the effects of small deviations in the transition matrix of a finite Markov chain.
J. Operations Res. Soc. Japan 16 (1973), 104 - 129.

[110] TAYLOR, H.: Markovian sequential replacement processes. Ann. Math. Statist. 36 (1965), 1677 - 1694.

[111] TEPLIZKY, M.G.: A possibility of a decomposition procedure of controlled semi-Markovian processes (Russ.).
Awt. i Telemech. 6 (1974), 14-22.

[112] TIJMS, H.C.: On dynamic programming with arbitrary state space, compact action space and the average return as criterion.
Preprint, Mathematisch Centrum, BW 55/75. Amsterdam 1975.

[113] VAN HEE, K.M.; HORDIJK, A.; VAN DER WAL, J.: Successive approximations for convergent dynamic programming.
In: Markov Decision Theory. Eds. H.C. Tijms & J. Wessels. Math. Centre Tracts No. 93, pp. 183 - 212. Amsterdam 1977.

[114] VAN NUNEN, J.A.E.E.: Contracting Markov Decision Processes. Math. Centre Tracts No. 71. Amsterdam 1976.

[115] VAN NUNEN, J.A.E.E.; WESSELS, J.: The generation of successive approximations for Markov decision processes by using stopping times.
In: Markov Decision Theory. Eds. H.C. Tijms & J. Wessels. Math. Centre Tracts No. 93, pp. 25 - 37. Amsterdam 1977.

[116] VEINOTT, A.F., Jr.: Discrete dynamic programming with sensitive discount optimality criteria.
Ann. Math. Statist. 40 (1969), 1635 - 1660.

[117] VEINOTT, A.F., Jr.: On finding optimal policies in discrete dynamic programming with no discounting.
Ann. Math. Statist. 37 (1966), 1284 - 1294.

[118] VISKOV, O.W.; SHIRJAJEV, A.N.: Policies leading to optimal stationary strategies.
Trudy Mat. Inst. V.A. Steklov 71 (1964), 35 - 45.

[119] WESSELS, J.: Markov programming by successive approximations with respect to weighted supremum norms.
J. Math. Anal. Appl. 58 (1977), 326 - 335.

[120] WHITE, D.J.: Dynamic programming, Markov chains and the method of successive approximations.
J. Math. Anal. Appl. 6 (1963), 373 - 376.

[121] WIJNGAARD, J.: Recurrence conditions and the existence of average optimal strategies for inventory problems on a countable state space.
In: Dynamische Optimierung. Bonner Math. Schriften Nr. 98, pp. 149 - 161. Bonn 1977.

[122] WIJNGAARD, J.: Sensitive optimality in stationary Markov decision problems on a general state space.
In: Markov Decision Theory. Eds. H.C. Tijms & J. Wessels.
Math. Centre Tracts No. 93, pp. 85 - 93. Amsterdam 1977.

[123] WIJNGAARD, J.: Stationary Markovian decision problems.
Discrete time, general state space.
Diss. Technische Hogeschool Eindhoven, 1975.

[124] WIJNGAARD, J.: Stationary Markovian decision problems and perturbation theory of quasi-compact linear operators.
Mathematics of Operations Research 2 (1977), 91 - 102.

[125] YAMADA, K.: Duality theorem in Markovian decision problems.
J. Math. Anal. Appl. 50 (1975), 579 - 595.

[126] ZESSIN, H.N.: Kategorientheoretische Behandlung der Zustandsraumtransformation von Markoffprozessen.
J. Reine Angew. Math. 260 (1973), 84 - 102.

SYMBOLS

1. GENERAL NOTATIONS

$o(.)$, $O(.)$	Landau symbols
$\mathbb{N}$	set of natural numbers
$\mathbb{Z}$	set of integers
$\mathbb{R}$	set of real numbers
$\mathbb{C}$	set of complex numbers
$\mathbb{N}_i$ $(i = -2, -1, 0) := \{N \in \mathbb{Z} : n \geq i\}$	
1_A	characteristic function of a set A
$\mathbb{1}$	1_E (E - state space)
$\preccurlyeq$	lexicographical order (p. 159)
$(E')^E$	set of all mappings $E \to E'$
$g \circ f$	$(g \circ f)(.) = g(f(.))$
$f\|_A = f_A$	restriction of a function to a set A
$\mathcal{B}_{\mathbb{C}}$	system of Borel sets of complex numbers
$B(E, \mathfrak{E}) = B(E)$	set of all $\mathfrak{E}$-measurable complex functions on E

2. SPECIAL NOTATIONS

(These symbols are defined in the text at the pages given here)

INDEX

MATHEMATICAL RESEARCH - MATHEMATISCHE FORSCHUNG

Band 1

INVERSE AND IMPROPERLY POSED PROBLEMS IN DIFFERENTIAL EQUATIONS

Proceedings of the Conference on Mathematical and Numerical Methods, held in Halle/Saale (GDR) from May 29 to June 2, 1979
Edited by Gottfried Anger
1979. 296 Seiten - 18 Abb. - 5 Tab. - gr. 8° - 37,-- M
Bestell-Nr. 762 744 1 (2182/1) in englischer Sprache

Band 2

FUNDAMENTALS OF COMPUTATION THEORY FCT '79

Proceedings of the Conference on Algebraic, Arithmetic and Categorial Methods in Computation Theory, held in Berlin/Wendisch-Rietz (GDR) September 17-21, 1979
Edited by Lothar Budach
1979. 576 Seiten - 27 Abb. - 2 Tab. - gr. 8° - 72,-- M
Bestell-Nr. 762 806 5 (2182/2) in englischer Sprache

Band 3

DECIDABILITY AND GENERALIZED QUANTIFIERS

By Andreas Baudiscn, Detlef Seese, Hans-Peter Tuschik, and Martin Weese
1980. XII, 235 Seiten - 16 Abb. - gr. 8° - 31,-- M
Bestell-Nr. 762 808 1 (2182/3) in englischer Sprache

Band 4

ZUR ELASTIZITÄTS- und THERMOELASTIZITÄTSTHEORIE

Potentialtheoretische Methoden in Anwendung auf verschiedene Modelle
Von Lothar Jentsch und Johannes Maul
1980. 157 Seiten - gr. 8° - 20,-- M
Bestell-Nr. 762 810 2 (2182/4)

Band 5

SYSTEMS ANALYSIS AND SIMULATION 1980

Proceedings of the International Symposium held at Berlin (GDR) September 1-5, 1980
Edited by Achim Sydow
1980. 496 Seiten - 111 Abb. - 11 Tab. - gr. 8° - 62,-- M;
Ausland 96,-- M - Bestell-Nr. 762 811 0 (2182/5) in englischer Sprache

Band 6

NICHTLINEARE DISKRETE OPTIMIERUNG

Von Eberhard Girlich und M. M. Kowaljow
1981. 218 Seiten - gr. 8° - 28,-- M - Bestell-Nr. 762 925 2 (2182/6)

Band 7

COMPLETENESS PROBLEMS IN THE STRUCTURAL THEORY OF AUTOMATA

By Jürgen Dassow
1981. 148 Seiten - gr. 8° - 20,-- M - Bestell-Nr. 762 951 9 (2182/7) in englischer Sprache

Band 8

DARSTELLUNG VON LÖSUNGEN LINEARER ELLIPTISCHER DIFFERENTIALGLEICHUNGEN

Von Günther Wildenhain
1981. 92 Seiten - gr. 8° - 12,-- M - Bestell-Nr. 763 003 4 (2182/8)

Band 9

INTRODUCTION TO THE SPECTRAL THEORY OF OPERATORS IN SPACES WITH AN INDEFINITE METRIC

By I. S. Iohvidov, M. G. Krein, and H. Langer
1982. 120 Seiten - gr. 8° - 15,-- M - Bestell-Nr. 763 041 3 (2182/9)
in englischer Sprache

Band 10

NONLINEAR TIME-DISCRETE SYSTEMS

A General Approach by Nonlinear Proposition
By Michael Gössel
1982. 116 Seiten - gr. 8° - 15,-- M - Bestell-Nr. 763 076 3 (2182/10)
in englischer Sprache

Band 11

PREPRIMAL ALGEBRAS

By Klaus Denecke
1982. 164 Seiten - gr. 8° - 22,-- M - Bestell-Nr. 763 103 7 (2182/11)
in englischer Sprache

Band 12

ERGODIC THEORY AND RELATED TOPICS

Proceedings of the Conference held in Vitte/Hiddensee (GDR)
October 19-23, 1981
Edited by Horst Michel
1982. 232 Seiten - gr. 8° - 30,-- M - Bestell-Nr. 763 138 7 (2182/12)
in englischer Sprache

Band 13

SINGULARLY PERTURBED DIFFERENTIAL EQUATIONS

By Herbert Goering, Andreas Felgenhauer, Gert Lube, Hans-Görg Roos, and Lutz Tobiska
1983. 176 Seiten - gr. 8° - 22,-- M - Bestell-Nr. 763 163 5 (2182/13)
in englischer Sprache

Band 14

LEBEN WIR IN EINER VOLTERRA-WELT?

Ein ökologischer Zugang zur angewandten Systemanalyse
Von Manfred Peschel und Werner Mende
1983. 234 Seiten - gr. 8° - 30,-- M - Bestell-Nr. 763 223 2 (2182/14)

Bezugsmöglichkeiten

Unsere Veröffentlichungen sind über die Buchhandlungen des In- und Auslandes erhältlich.

Bitte wenden Sie sich

im sozialistischen Ausland	an die Buchhandlungen für fremdsprachige Literatur
in der BRD und in Berlin (West)	an KUNST UND WISSEN, Erich Bieber, Wilhelmstraße 4-7, D-7000 Suttgart 1
bei Wohnsitz im übrigen Ausland	an eine internationale Buchhandlung oder an den BUCHEXPORT, Volkseigener Außenhandelsbetrieb der DDR, DDR-7010 Leipzig, Leninstraße 16

Bei Fragen zur Produktsicherheit wenden Sie sich bitte an:
If you have any questions regarding product safety,
please contact:

Walter de Gruyter GmbH
Genthiner Straße 13
10785 Berlin
productsafety@degruyterbrill.com